Computational Intelligence and Deep Learning Methods for Neuro-rehabilitation Applications

Computational Intelligence and Deep Learning Methods for Neuro-rehabilitation Applications

Edited by

D. Jude Hemanth

Professor, ECE Department, Karunya Institute of Technology and Sciences, Coimbatore, India

ELSEVIER

ACADEMIC PRESS

An imprint of Elsevier

Academic Press is an imprint of Elsevier
125 London Wall, London EC2Y 5AS, United Kingdom
525 B Street, Suite 1650, San Diego, CA 92101, United States
50 Hampshire Street, 5th Floor, Cambridge, MA 02139, United States
The Boulevard, Langford Lane, Kidlington, Oxford OX5 1GB, United Kingdom

ISBN: 978-0-443-13772-3

For information on all Academic Press publications visit our website at https://www.elsevier.com/books-and-journals

Publisher: Mara Conner
Acquisitions Editor: Carrie Bolger
Editorial Project Manager: Emily Thomson
Production Project Manager: Erragounta Saibabu Rao
Cover Designer: Vicky Pearson Esser

Typeset by TNQ Technologies

Contents

Contributors

Srinath Akuthota
Department of Electronics and Communication Engineering, SR University, Warangal, Telangana, India

A. Ananthi
Department of Robotics Engineering; Karunya Institute of Technology and Sciences, Coimbatore, Tamil Nadu, India

J. Anitha
Department of Electronics and Communication Engineering, Karunya Institute of Technology and Sciences, Coimbatore, Tamil Nadu, India

D. Abraham Chandy
Karunya Institute of Technology and Sciences, Coimbatore, Tamil Nadu, India

A. Hepzibah Christinal
Karunya Institute of Technology and Sciences, Coimbatore, Tamil Nadu, India

Kanav Dhir
Department of Chemistry, S.C.D Government College, Ludhiana, Punjab, India

S. Thomas George
Department of Biomedical Engineering; Karunya Institute of Technology and Sciences, Coimbatore, Tamil Nadu, India

Deepika Ghai
School of Electronics & Electrical Engineering, Lovely Professional University, Phagwara, Punjab, India

Shivalika Goyal
Imaging, Avionics and Display Systems Department, CSIR-CSIO, Chandigarh, Punjab, India

Rajeev Gupta
Department of Computer Science and Engineering, M.M. Engineering College, Maharishi Markandeshwar (Deemed to be University), Mullana, Ambala, India

Ravichander Janapati
Department of Electronics and Communication Engineering, SR University, Warangal, Telangana, India

A. Jenefa
Department of Computer Science and Engineering, Karunya Institute of Technology and Sciences, Coimbatore, Tamil Nadu, India

Dilan Kalaycı Alas
Atatürk Faculty of Education, Near East University, Northern Cyprus, Nicosia, Turkey

S. Karthic
Department of Computer Science and Engineering, KPR Institute of Engineering and Technology, Coimbatore, Tamil Nadu, India

K. Kathiresan
Department of Computer Science and Engineering, KPR Institute of Engineering and Technology, Coimbatore, Tamil Nadu, India

Ranjit Kaur
School of Computer Applications, Lovely Professional University, Phagwara, Punjab, India

K. Raj Kumar
Department of Electronics and Communication Engineering, SR University, Warangal, Telangana, India

Amit Laddi
Imaging, Avionics and Display Systems Department, CSIR-CSIO, Chandigarh, Punjab, India

A. Lincy
Department of Computer Science and Engineering, National Engineering College, Kovilpatti, Tamil Nadu, India

P. Manimegalai
Department of Biomedical Engineering; Karunya Institute of Technology and Sciences, Coimbatore, Tamil Nadu, India

Mihnea Ion Marin
University of Craiova, Craiova, Romania

K. Mouthami
Department of Computer Science and Engineering, KPR Institute of Engineering and Technology, Coimbatore, Tamil Nadu, India

Vaishnavi Naveen
Department of Electronics and Communication Engineering, Dayananda Sagar Academy of Technology and Management, Bangalore, Karnataka, India

Behçet Öznacar
Atatürk Faculty of Education, Near East University, Northern Cyprus, Nicosia, Turkey

J. Prasanna
Department of Biomedical Engineering; Karunya Institute of Technology and Sciences, Coimbatore, Tamil Nadu, India

T. Preethi
Department of Computer Science and Engineering, KPR Institute of Engineering and Technology, Coimbatore, Tamil Nadu, India

Kirti Rawal
School of Electronics & Electrical Engineering, Lovely Professional University, Phagwara, Punjab, India

C. Alvino Rock
Karunya Institute of Technology and Sciences, Coimbatore, Tamil Nadu, India

Ligia Rusu
University of Craiova, Craiova, Romania

N.J. Sairamya
Department of Psychology; University of Calgary, Calgary, Canada

Ramandeep Sandhu
School of Computer Science & Engineering, Lovely Professional University, Phagwara, Punjab, India

S.A. Srinivasan
Department of Computer Science and Engineering, KPR Institute of Engineering and Technology, Coimbatore, Tamil Nadu, India

K.R. Sri Preethaa
Department of Computer Science and Engineering, KPR Institute of Engineering and Technology, Coimbatore, Tamil Nadu, India; Department of Robot and Smart System Engineering, Kyungpook National University, Daegu, Republic of Korea

M.S.P. Subathra
Department of Robotics Engineering; Karunya Institute of Technology and Sciences, Coimbatore, Tamil Nadu, India

M.N. Sumaiya
Department of Electronics and Communication Engineering, Dayananda Sagar Academy of Technology and Management, Bangalore, Karnataka, India

Srivatsa Sundarraj
Department of Computer Science and Engineering, KPR Institute of Engineering and Technology, Coimbatore, Tamil Nadu, India

Suman Lata Tripathi
School of Electronics & Electrical Engineering, Lovely Professional University, Phagwara, Punjab, India

G.S. Vachanamruth
Department of Electronics and Communication Engineering, Dayananda Sagar Academy of Technology and Management, Bangalore, Karnataka, India

C. Varshitha
Department of Electronics and Communication Engineering, Dayananda Sagar Academy of Technology and Management, Bangalore, Karnataka, India

Gitanjali Wadhwa
Department of Computer Science and Engineering, KPR Institute of Engineering and Technology, Coimbatore, Tamil Nadu, India

V.P. Yashaswini
Department of Electronics and Communication Engineering, Dayananda Sagar Academy of Technology and Management, Bangalore, Karnataka, India

N. Yuvaraj
Department of Computer Science and Engineering, KPR Institute of Engineering and Technology, Coimbatore, Tamil Nadu, India

Preface

Neurological disorders are a significant area of concern, not only for the neuro patients but also for the elderly. Several therapeutic treatments are continuously being made available for neuro-disorders. However, the success rate is limited. An alternate/add-on for the therapeutic treatment is the neuro-rehabilitation process which is booming in the current scenario. With the vast improvement in the quality of life of the neuro patients, rehabilitation methods coupled with technology are on the rise in the medical sector. However, the integration of advanced technologies such as Artificial Intelligence (AI) with neuro-rehabilitation is not yet fully discovered. The main objective of the contents of this book is to throw more insights on AI-based neuro-rehabilitation methods. This book is an interdisciplinary book which suits the health professionals and the computer scientists. The brief overview of the different chapters is given in the subsequent paragraphs.

Chapter 1 is an introduction chapter which portrays the different types of AI-based technologies available for neuro-rehabilitation applications. This chapter provides the list of computer apps available for neuro-rehabilitation. The pros and cons of the several methods are also highlighted in this chapter. Chapter 2 deals with statistical and computational approaches for gait assessment which is necessary for neuro-rehabilitation. The muscular movements are assessed using the computer-aided methods which decide on the success rate of the overall neuro-rehabilitation process. Remote monitoring of neuro patients is the focal point of Chapter 3. Deep Learning (DL) methods are used to assess the improvement in the patients. Dementia/Alzheimer is the focal point of this study.

Chapter 4 deals with rehabilitation approaches for patients suffering from autism disorder. Mixed reality is the prime technology used in the experiments. Wearable sleeves for assessing the quality of life in neuro patients are explored in Chapter 5. This technology is used in physiotherapy applications. Different sensors and Graphical User Interface (GUI) are the significant constituents of these experiments. Assistive devices are an integral part of neuro-rehabilitation. The role of Machine Learning (ML) in developing these assistive devices is explored in Chapter 6.

Chapter 7 deals with ML and DL approaches to improve the quality of life of patients with speech and language disorders. Classical Artificial Neural Networks (ANNs) are used in these experiments to analyze the different types of disorders such as stuttering, articulation disorder, etc. Patients with trauma injuries are the focus of Chapter 8. Several ML approaches are used in these experiments to assess the cognitive ability and subsequent treatment planning. Brain Computer Interface (BCI) is another allied area of neuro-rehabilitation which is dealt in Chapter 9. The accuracy of the data such as EEG is very important to train and test the AI-based algorithms in the computer. However, the acquired EEG signals are mostly noisy which prevents the practical feasibility of the proposed approaches. This chapter concentrates on developing algorithms to eliminate the noise in source data to ensure the success rate of the overall BCI system.

Neuro-rehabilitation for quadriplegic patients with the support of BCI is illustrated in Chapter 10. Convolution Neural Networks (CNNs) are used in this work for the experiments. An extensive analysis is carried out to show the superior nature of the proposed method. The concepts of motor imagery and DL are used in the study to classify the different EEG signals in Chapter 11. A hybrid wavelet transform–based DL model is used in this work for EEG signal classification. A complete neuroscience-based approach is explored in Chapter 12. The neuron activities in the brain responsible for the various disorders are studied in this work. Few modeling concepts are also illustrated in this chapter.

I am thankful to the contributors and reviewers for their excellent contributions in this book. My special thanks to Elsevier, especially to Ms. Carrie Bolger (Acquisition Editor) for the excellent collaboration. Finally, I would like to thank Ms. Emily Thomson who coordinated the entire proceedings. This edited book covers the fundamental concepts and application areas in detail. Being an interdisciplinary book, I hope it will be useful for both health professionals and computer scientists.

Dr. D Jude Hemanth
April, 2023

CHAPTER 1

AI-based technologies, challenges, and solutions for neurorehabilitation: A systematic mapping

Rajeev Gupta

Department of Computer Science and Engineering, M.M. Engineering College, Maharishi Markandeshwar (Deemed to be University), Mullana, Ambala, India

1. Introduction

A wide range of businesses, particularly the healthcare sector, have been significantly impacted by artificial intelligence (AI). This cutting-edge technology is no longer just a pipe dream. Instead, this rapidly developing technology has impacted day-to-day life, which cannot be predicted by a human being. Artificial intelligence applications in healthcare are essentially changing the information technology era.

1.1 Artificial intelligence

The capacity to acquire and use knowledge and skills is what is meant by intelligence. Artificial intelligence is the ability displayed by machines to execute a variety of tasks with the aid of sentiment analysis and natural language processing (NLP). It is a group of intelligent operations and actions created by technology and computer models. With the use of this technology, machines are now capable of understanding the information they are given and using it to carry out a variety of business functions. Machine learning and deep learning, which are subsets of AI, each has distinct roles to play in the training of machines. Humans have intelligent effectors, including the mouth, upper and lower limbs, and other bodily components that carry out commands.

Robotic agents and infrared range seekers are used as effectors in robotics and cameras [1]. To provide near-precision data with better clinical evidence for various medical diagnoses, treatment options, and decisions, including medical research and a wide spectrum of the entity of the developing health care system, computers' usage has been optimized, and their access to enormous data from a wide range of advanced computations models and algorithms has increased [1]. To increase the effectiveness of artificial intelligence in the healthcare business, health organizations have started partnerships with several technological firms. It is now widely employed in several medical applications, including the treatment, diagnosis,

Computational Intelligence and Deep Learning Methods for Neuro-rehabilitation Applications
https://doi.org/10.1016/B978-0-443-13772-3.00014-5

prognosis, and rehabilitation of diseases. It has been demonstrated that using intelligent algorithms and human intelligence together improves treatment.

1.2 Neurological disorders

Neurological illnesses are triggered by the amalgamation of structural abnormalities with electrical and biochemical abnormalities in the nervous systems of a human being. These conditions might be hereditary, progressive, or severe in nature. Few of these are generated by surgical procedures, but the vast majority require pharmaceutical intervention and maintenance and are likely to worsen. These illnesses generally affect mental functions. Neurological disorders are diseases that are related to the spinal cord and brain. These disorders are also leaving behind their mark on neuromuscular junctions, muscles, roots of nerves, and other nerves like peripheral, cranial nerves, and autonomic. Neurological illnesses are the major cause of disability globally, according to current statistics, and their contributions to the overall burden are growing [2].

Because of their growing and aging populations, underdeveloping and developing countries tolerate the puma stake of this conjunction. As a result, there is a growing demand for neurological disorders treatment, rehabilitation, and support services. Because of the high complexity of the universal conjunction of neurological disorders, there are numerous unsolved issues in the domains of proper management, early diagnosis, and assured solution, which are exacerbated in the areas where insufficient medical services are available. According to experts, there are currently insufficient modifiable risks for these disorders, and to create effective preventative and treatment techniques, fresh understanding is necessary.

In addition to the neurological examination in clinical practice, which produces precise symptoms and signs, specific tests from different techniques are employed to arrive at a functional diagnosis. The neurologist's clinical reasoning process ensures the accuracy of the diagnosis despite the availability of these technologies. For functional diagnosis, various tests are conducted, and based on these tests, an appropriate diagnosis is conducted. The tests prescribed by clinical practitioners include laboratory screening tests, genetic testing, and brain scans (which consist of computed tomography (CT) scan, single proton emission (SPECT) scan, and magnetic resonance imaging (MRI) scan). In addition, angiography, biopsy, cerebrospinal fluid analysis, electroencephalography (EEG), electromyography (EMG), and electronystagmography (ENG) evoked potentials, myelography, polysomnogram, thermography, ultrasound imaging, and X-rays.

A method to lessen neurological abnormalities under the supervision of medical professionals is called neurorehabilitation. Recent advancements in neurorehabilitation and AI techniques have increased our understanding of brain networks, allowed for the early diagnosis of neurological problems, and helped patients do their jobs with efficiency. Neuroscientists are now able to look at how the brain functions and identify the causes of abnormalities in neurological conditions such as epilepsy, neuromuscular diseases, autism, attention deficit disorder (ADD), brain tumors, dementia, Parkinson's disease, and cerebral palsy. These advancements enable strategies for cognitive rehabilitation, prognostic marker discovery, and early and accurate diagnosis.

The common neurological conditions or ailments of the nervous system mentioned in the International Classification of Diseases, ICD-11 document are displayed in Table 1.1. A universal standard for diagnostic health information, clinical recording, and statistical aggregation is depicted by ICD-11. This ICD version, which has the goal of providing primary care, is multipurpose, completely updated,

Table 1.1 Common neurological disorders mentioned in the ICD-11 document [3].

Movement disorders
Disorders with neurocognitive impairment as a major feature
Multiple sclerosis or other white matter disorders
Epilepsy or seizures
Headache disorders
Cerebrovascular diseases
Spinal cord disorders excluding trauma
Motor neuron diseases or related disorders
Disorders of the nerve root, plexus, or peripheral nerves
Diseases of neuromuscular junction or muscle
Cerebral palsy
Nutritional or toxic disorders of the nervous system
Disorders of cerebrospinal fluid pressure or flow
Disorders of the autonomic nervous system
Human prion diseases
Disorders of consciousness
Other disorders of the nervous system
Postprocedural disorders of the nervous system
Injuries to the nervous system
Neoplasms of the nervous system
Structural developmental anomalies of the nervous system
LD20 syndromes with central nervous system anomalies as a major feature
Nonviral and unspecified infections of the central nervous system
Symptoms, signs, or clinical findings of the nervous system
Paralytic symptoms
6B60 Dissociative neurological symptom disorder
JB64.3 Diseases of the nervous system complicating pregnancy, childbirth, or the puerperium
8E7Y Other specified diseases of the nervous system
8E7Z Diseases of the nervous system, unspecified

scientifically modern, and created for use in a digital environment. The international standard for compiling, reporting, analyzing, interpreting, and comparing statistics on mortality and morbidity is called ICD-11. This 11th iteration, which makes it useable by both these groups and coders, is the result of an unparalleled collaboration with physicians, statisticians, categorization, and IT specialists from around the world.

The ICD-11 document lists all common and important health issues. Each individual or group of common health issues are having unique ICD-11 codes. These codes are producing data that can be used for clinical documentation or by governments to plan and evaluate effective public health policies. The electronic version of ICD documents consists of more than 17,000 diagnostic categories and approximately 100,000 medical phrases for medical analytical indexing. Both online as well as offline versions are available of ICD-11 for medical practitioners.

The author used a narrative review approach in this chapter, which provides a more wide-ranging way comparison of regular reviews to analyze and represent the eminent researcher's views in the field of artificial intelligence and neurorehabilitation.

Section 2 of this chapter discusses the recent developments and involvement of Artificial Intelligence in Healthcare and Neurorehabilitation. Section 3 discusses the major clinical challenges and robotic rehabilitation applications. Section 4 is concerned with technological synergies driving neural rehabilitation. Section 5 discusses the procedure for selecting AI-based supportive technologies and tools for initiating neurorehabilitation. Finally, the chapter conclusion and future scope are illustrated in Section 6.

2. Recent developments and involvement of artificial intelligence in healthcare and neurorehabilitation

In today's digital era, prominent advancement in the arena of neurorehabilitation is made possible by high-tech revolution and amended neuro-physiological understanding. These advancements are moving ahead toward new research milestones. These eminent advancements make existing interventions more relevant while also allowing for novel and unique methodologies and high-tech collaborations. New approaches use either invasive medical mechanisms or noninvasive medical mechanisms-based human–machine interfaces to help people overwhelmed by long-term diseases or disabilities due to injury of the spinal cord. People affected with Duchenne muscular dystrophy were overcome by electromyography and novel sensors using neuro-motor mechanisms. People, who are suffering from stroke, amputation, and traumatic brain injury are overwhelmed by using wearable robotics devices.

Furthermore, the motor-learning skills of a patient may improve by enhancing robot-assisted rehabilitation and also produce movement recurrences by interpreting patients' brain motion through remedy. The combination of artificial intelligence techniques with advanced digital machinery can be further assisted. Such mentioned technologies with integrations of neurorehabilitation action can have a significant practical impact. These technological changes have the potential to enable and motivate nontechnically persons including healthcare supporting staff, family members of a patient, and others, to change the technical knowledge of neural-rehabilitation robotic tools. With the help of such type of knowledge enhancing and allowing them to enthusiastically participate in the care and support of the disabled person. This narrative review looks at current and evolving neurorehabilitation techniques in terms of encouraging or enlisting dormant neurorehabilitation, enhancing or improving natural neural output, and changing or reinstating functions.

According to ChubbyBrain (CB) Insights, AI sector funding is expected to increase by approximately 156% by 2025, with healthcare accounting for roughly a fifth of total funding. Artificial intelligence is reinventing and reviving the domain of modern digital healthcare via enhanced electronic machines. With the help of this uplifted technology, a clinical practitioner or supportive healthcare staff can understand, study, analyze, apply, and act on the technological AI mechanisms in neurorehabilitation. Further, these technological improvements are used to discover new associations among hereditary codes and operation-theater surgery-assisted medical robots. Fig. 1.1 illustrates the involvement of artificial intelligence in healthcare and neurorehabilitation.

Artificial intelligence and its subdomains are efficiently used to access and deploy accurate inventions in the healthcare sector. These inventions will be

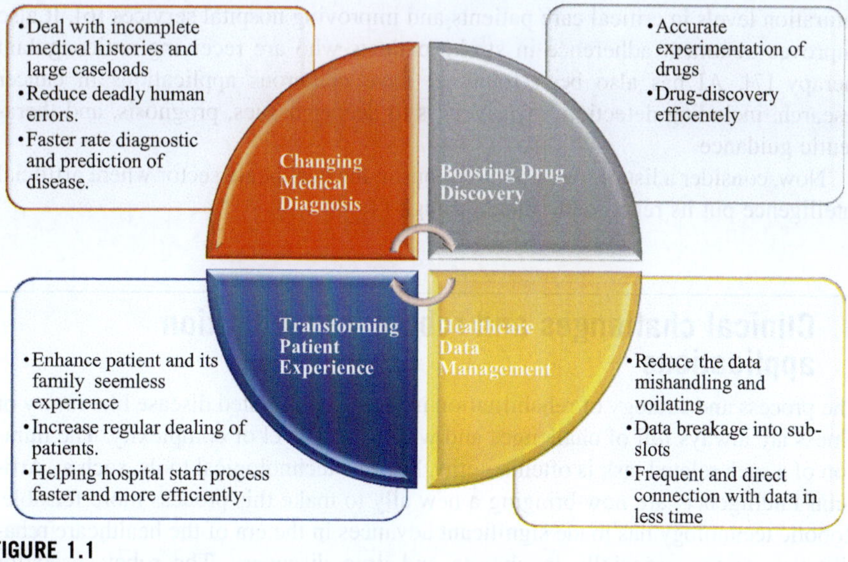

FIGURE 1.1

Involvement of artificial intelligence in healthcare and neurorehabilitation.

providing support and take care of chronic diseases suffered by patients and hopefully, also help to cure them speedily. With the concern of clinical decision-making and healthcare business analytics, artificial intelligence has lots of strength as compared to existing nontechnological traditional mechanisms. AI techniques expand healthcare system accuracy with the help of a proper understanding of training as well as testing datasets, and appropriate solutions to gain unique visions into treatment inconsistency, In-patient department (IPD) care progressions, proper diagnostics, and patient-recovery outcomes. Acumen Research study predicts that by the end of 2026, the international business of Artificial Intelligence in the healthcare industry will achieve a target of up to $8 billion.

With a plethora of AI tools and platforms now available, many health organizations have begun collaborating with technology companies to improve the use of AI in the health care entity [4]. Artificial intelligence is widely used in medical applications such as disease management and screening, diagnosis, prognosis estimation, and therapy. Artificial intelligence has just acknowledged a lot of consideration due to its benefits in the treatment of cancer, diabetes, coronary artery disease, stroke, and other neurodegenerative diseases [5]. Patient monitoring, guidance, and status assessment are all important applications of AI. This has also resulted in AI applications in areas such as mobile computing, intending to increase creativity and provide better disease management services.

It is also known that AI in the medical domain has an extensive assortment of technological applications in chronic respiratory disease with lifestyle management. To manage and control various chronic conditions, AI systems integrate geographical and clinical data with sensor-based computational technology in the form of various wearable devices [4]. Artificial intelligence is useful in monitoring oxygen saturation levels in critical care patients and improving hospital services [6]. It also improves treatment adherence in stroke patients who are receiving anticoagulant therapy [7]. AI has also been found to have numerous applications in cancer research, including detection, invasiveness to deeper tissues, prognosis, and therapeutic guidance.

Now, consider a list of major applications in the healthcare sector where artificial intelligence put its remarkable breakthrough (Table 1.2).

3. Clinical challenges and robotic rehabilitation applications

The process and strategy of rehabilitation in any health-related disease like injury or illness are always full of challenges and with a high level of complexity. The duration of a such related task is often lengthy, but new technological tools, such as Artificial Intelligence, are now bringing a new ally to make this process more feasible. Robotic technology has made significant advances in the era of the healthcare rehabilitation sector, especially in pharma and drug discovery. The robots used for

Table 1.2 Major applications of artificial intelligence in healthcare sector.

Artificial intelligence—changing medical diagnosis	
App name	**Description**
	App name: Viz.ai **Purpose and benefits:** • Provides AI-powered healthcare faster treatment • Quick decision power for saving lives • Detect unavoidable health-related issues and care for them quickly **Site**: San Francisco, California
	App name: PathAI **Purpose and benefits:** • Develops machine-learning-based technology for assisting the pathologists • Reducing errors in cancer diagnosis • Worked with drug developers to expand AI-based technology in the healthcare industry. **Site**: Boston, Massachusetts
	App name: Regard **Purpose and benefits:** • Diagnose patients • Receive specific recommendations about patient care. • Automatically updates patient's health-related documents **Site**: Los Angeles, California
	App name: Buoy Health **Purpose and benefits:** • Diagnose patients and treat illness by AI-based symptom and cure checker • Helps in diagnosis and quick treatment of patients. **Site**: Boston, Massachusetts
	App name: Enlitic **Purpose and benefits:** • Diagnose radiology patients • Uses deep-learning medical tools and handle unstructured medical data • Helps in diagnosis and quick treatment of patients. **Site**: San Francisco, California
	App name: Freenome **Purpose and benefits:** • Used in diagnostic test and screenings • Handle blood work for cancer patients. • Detect cancer in its earliest stages **Site**: San Francisco, California
	App name: Beth Israel Deaconess Medical Center **Purpose and benefits:** • Potentially diagnose deadly blood diseases • Detect concerned diseases at the early stage of growth • Developed microscopes for harmful bacteria scanning **Site**: Boston, Massachusetts

Continued

Table 1.2 Major applications of artificial intelligence in healthcare sector.—*cont'd*

Artificial intelligence—changing medical diagnosis	
App name	**Description**
	App name: Iterative Health **Purpose and benefits:** • Potentially diagnose and handle gastroenterology diseases • Identifying concerned patients having diseases like inflammatory bowel • Developed SKOUT device for clinical trial **Site**: Cambridge, Massachusetts
	App name: VirtuSense **Purpose and benefits:** • Potentially diagnose and handle patient's movement • Identifying concerned patients having a risk of falling • Developed VSTAlert device for clinical trial **Site**: Peoria, Illinois
	App name: Caption Health **Purpose and benefits:** • Early diagnose and identification of disease. • Handle ultrasound process and produce quality images for further diagnostic. **Site**: Brisbane, California
Artificial intelligence—boosting drug discovery	
	Company name: BioXcel Therapeutics **Purpose and benefits:** • Identify and develop new medicines in the domain of immuno-oncology • Identify and develop new medicines in the domain of neuroscience. **Site**: New Haven, Connecticut
	Company name: Reverie Labs **Purpose and benefits:** • Connecting machine learning tools with computational chemistry for new drug design and discovery. • Identify and develop effective cancer treatments. • Used expansive databases and predictive analytics tools **Site**: Cambridge, Massachusetts
	Company name: Valo Health **Purpose and benefits:** • Collected human-centric data with the help of Opal Computation Platform to identify general diseases. • Established clinical development and the molecule design **Site**: Boston, Massachusetts
	App name: BERG **Purpose and benefits:** • Maps the diseases to accelerate the discovery and development of breakthrough medicines. • Launches new products for handling rare diseases. **Site**: Framingham, Massachusetts

Table 1.2 Major applications of artificial intelligence in healthcare sector.—*cont'd*

	Artificial intelligence—changing medical diagnosis
App name	**Description**
	App name: XtalPi **Purpose and benefits:** • Combinations of cloud, AI, and quantum physics and developed a product for further predictions of pharmaceutical and chemical properties of drugs. • Predicts complex molecular systems with the help of crystal structure prediction technology in less time **Site**: Cambridge, Massachusetts
	App name: Atomwise **Purpose and benefits:** • Tackle complex diseases like multiple sclerosis and Ebola. • Predicts bioactivity with the help of AtomNet and identifies patient characteristics. • Fast response app **Site**: San Francisco, California
	App name: Deep Genomics **Purpose and benefits:** • Identify and tackle the right candidates for drugs development related to neuro-degenerative and neuro-muscular disorders • Analyzes different cell compounds with the help of Project Saturn. **Site**: Fully Remote
	App name: BenevolentAI **Purpose and benefits:** • Provide a timely right treatment in accurate manner to a right needy patient. • Licensed drugs with the help of major pharma companies. • Developed easily transportable medicines for rare diseases.
Artificial intelligence—transforming patient experience	
	Company name: Iodine Software **Purpose and benefits:** • Developed AI-based software for mid-level revenue cycle leakages • Provide optimization of resources and enhance response rate • Discover document accuracy and clients' insights **Site**: Austin, Texas
	Company name: Kaia Health **Purpose and benefits:** • Developed and operated a digital therapeutics platform • Case reviews, routines of exercise, and other activities of relaxation are major highlights • Featured a PT-grade automatic coach for feedback **Site**: New York, New York

Continued

Table 1.2 Major applications of artificial intelligence in healthcare sector.—*cont'd*

Artificial intelligence—changing medical diagnosis	
App name	**Description**
	Company name: Spring Health **Purpose and benefits:** • Developed and offered software for mental health benefit • Tele-health appointment-based product **Site**: New York, New York
	Company name: Twin Health **Purpose and benefits:** • Developed and offered software products for handling Type-2 diabetes • Launched product (Whole body Digital Twin) is based on IoT, AI, medical science, and data science **Site**: Mountain View, California
	Company name: Olive **Purpose and benefits:** • Developed an automated repetitive healthcare tasks platform • Checks the eligibility to unadjudicated claims and data migrations • Eliminates the costly integrations **Site**: Columbus, Ohio
	Company name: Qventus, Inc. **Purpose and benefits:** • Developed and manage the software that solves operational challenges • Checks and manage patient's safety, injury, illness • Handle automated emergency room, ambulance routes, and patient waiting times. **Site**: Mountain View, California
	Company name: Cleveland Clinic **Purpose and benefits:** • Supports research in the domains of population health, chemicals, drug-discovery • Innovating patient care **Site**: Cleveland, Ohio
	Company name: Johns Hopkins Medicine **Purpose and benefits:** • Use predictive AI techniques for patient operational flow. • Manage emergency IPD and partnered with GE Healthcare **Site**: Baltimore, Maryland
	Company name: Babylon **Purpose and benefits:** • Handle patient by using a symptom checker • Monitor and help people who want to stay healthy **Site**: Austin, Texas

Table 1.2 Major applications of artificial intelligence in healthcare sector.—*cont'd*

Artificial intelligence—changing medical diagnosis	
App name	**Description**
	Company name: One Drop **Purpose and benefits:** • Provide discrete solutions for managing diabetes and high BP. • Manage and deal with weight and also, work as a glucose reader **Site**: New York, New York
	Company name: CloudMedx **Purpose and benefits:** • Provide support to hospitals and private clinics for managing patient data. • Manage payment information by implementing predictive analytics. **Site**: Palo Alto, California
	Company name: Subtle Medical **Purpose and benefits:** • With the help of AI, it enhances medical images for the radiology department. • Reduce image noise and speed up PET and MRI scans and also reducing scan time **Site**: Menlo Park, California
	App name: Twill **Purpose and benefits:** • Deliver personalized care programs • Provide facilities of care communities and digital therapeutics **Site**: New York, New York
Artificial intelligence—healthcare data management	
	Company name: Tempus **Purpose and benefits:** • Develop AI tools for physicians' insights for further patient cures and treatment. • Manage and collect clinical and molecular data for healthcare treatments **Site**: Chicago, Illinois
	Company name: IBM **Purpose and benefits:** • Develop personalized patient health plans for catching early signs of disease. **Site**: Armonk, New York
	App name: ClosedLoop.ai **Purpose and benefits:** • Discover high-risk patients and recommend treatment accordingly. • Used by Pharma companies, healthcare providers, payers, and life science companies. **Site**: Austin, Texas

Continued

Table 1.2 Major applications of artificial intelligence in healthcare sector.—*cont'd*

Artificial intelligence—changing medical diagnosis	
App name	**Description**
	App name: Beacon Biosignals **Purpose and benefits:** • Provide treatment to psychiatric and neurological diseases patients through EEG analytics. • Manage population stratification **Site**: Fully Remote
	App name: KenSci **Purpose and benefits:** • Identify and predict financial, clinical, and operational risk. **Site**: Seattle, Washington
	App name: Proscia **Purpose and benefits:** • Identify and detect group of patterns in cancer tissue. • Used image analytics by pathology labs for cancer discovery **Site**: Philadelphia, Pennsylvania
	App name: H2O.ai **Purpose and benefits:** • Identify and predict ICU transfers and, improve clinical workflows • Helps to mine health data for further detection of sepsis. **Site**: Mountain View, California
	App name: iCarbonX **Purpose and benefits:** • Identify and classify patients' symptoms for further treatment. **Site**: Shenzhen, China
	Company name: Akasa **Purpose and benefits:** • Helps healthcare providers for managing systematic workflows. **Site**: San Francisco, California
Artificial intelligence—enhancing robotic surgery	
	Company name: Vicarious Surgical **Purpose and benefits:** • Helps surgeons for minimal invasive operations • Virtually shrink and explore the patient's body internally. **Site**: Waltham, Massachusetts
	Company name: Auris Health **Purpose and benefits:** • Helps to improve and enhance endoscopies • Virtually study lung cancer with its treatment **Site**: Redwood City, California

Table 1.2 Major applications of artificial intelligence in healthcare sector.—*cont'd*

	Artificial intelligence—changing medical diagnosis
App name	**Description**
	Company name: Accuracy **Purpose and benefits:** • Helps and treat precisely cancerous tumors **Site**: Sunnyvale, California
	Company name: Intuitive **Purpose and benefits:** • Captured information quickly and provide analytics to surgeons for improved surgery. **Site**: Sunnyvale, California
	Company name: Microsure **Purpose and benefits:** • Captured information quickly and improve performance and provide precise information to surgeons for improved surgery. **Site**: Eindhoven, Netherlands

rehabilitation purposes are useful due to the technological advancement and patients' speedy recovery. Each therapy session is captured and also recorded by the robots. The captured or recorded data may be used for research purposes or to track the patient's progress. It can also be used to encourage patients to further health progress.

3.1 Clinical challenges

The major challenges faced by practitioners for rehabilitation and clinical practice are speckled here.

- *Safety:* Multiple patients are treated in the same room as in the past, and because of COVID-19 variants, it is difficult to maintain by adhering to stringent and mandatory safety measures.
- *Financial stability:* Many of us are worried about how the economy will change in 2023. A systematic planning and optimizing strategy present a significant challenge for physiotherapists and other practitioners.
- *Speeding up of recovery:* Getting the patient back to work as soon as possible is crucial for those of us who work in the field of physiotherapy and rehabilitation. When it comes to integrating digital tools into clinical practice, healthcare professionals need to be on the cutting edge.
- *Differentiation:* Due to easy access to information, patients are becoming more and more interested in better performance. They look for treatments that are personalized for them, improve their health outcomes, and guarantee the best

possible care in a secure setting. Today, this difficulty offers a genuine chance. The World Health Organization (WHO) regards digital health solutions as one of the most promising means of addressing this issue in contemporary societies.

- *Treatment customization:* There is nothing novel about the need for treatments to be tailored to the patient rather than the other way around. When patients receive specialized care, they develop differently and more effectively. The issue is typically a lack of time to create these plans and implement them one at a time.
- *Patient's progress:* The ongoing challenge is to improve the patient. It is obvious that it is not something special for 2023, but we have chosen to save it for last because it is always the most crucial. Of course, it's a complex problem, and as clinicians, it depends on us.

Fig. 1.2 shows the major challenges faced by practitioners for rehabilitation and clinical practice.

In addition, the following neurological disorders are critical challenges for a practitioner:

Stroke is defined as an interruption during the supply and drainage of blood to and from the brain, or in other sense, it is an interruption during the inbox and outbox

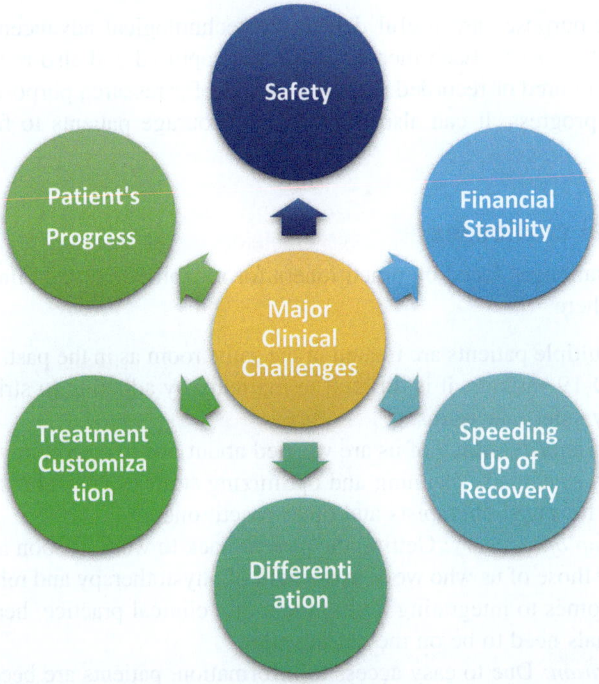

FIGURE 1.2

Major challenges faced by practitioners for rehabilitation and clinical practice.

of blood to or from the blood vessel wall of the brain, which results in the damaging of tissues of the brain. Robotics can be used alone or in conjunction with traditional rehabilitation to provide effective neurorehabilitation. The usage of neurorobotics in the healthcare sector or rehabilitation achieves its target due to their strong ability of continuous rehabilitation performance, rehabilitation's intensity adjustment, usage of customized self-constrained training setup, and reduction of physical burden on physiotherapist's physical burden. Rehabilitation robots, in conjunction with clinical's old-fashioned therapy, can provide concentrated working out with positive outcomes.

Every year, nearly 10 million people worldwide suffer from continuing or permanent disability due to hereditary, bad food habits, or nonintentionally reasons and on this point, the need for neurorehabilitation is required. Neurorehabilitation has been investigated for the treatment of these mild to moderate diseases, as well as for enhancing the quality of life. Devotion, shyness, and remembrance are all cognitive functions that can be targeted by robotic neurorehabilitation.

Motor vehicle accidents and falls account for roughly 80%–90% incidents due to spinal cord injury (SCI). Although the average age of injury has allegedly climbed to 42–50 years in recent decades, this ailment mostly affects young male individuals. In the form of robotic-assisted gait training, bodyweight support treadmill systems, and exoskeletons [8], robotics are now a mainstay of rehabilitation therapies for SCI sufferers [2], particularly for patients with partial paralysis.

Prosthetics and bionics are frequently used to assist patients who have lost limbs. We must provide quick prosthetic fitting, effective rehabilitation, and posttraumatic therapy to achieve the optimum prosthetic benefit for traumatic amputees. Significant progress has been achieved in creating prosthetic devices targeted at restoring human capabilities to persons who have lost one or more limbs in recent years. Some of the developments that allow more amputees to live independently include power and electrical controls, socket construction and fitting procedures, components, and suspension systems. Upper-limb amputee rehabilitation has benefited from technology advancements such as myoelectric and proportionately controlled devices, as well as elbow joints.

Some of the advancements presently employed for lower limb amputees include ankle rotators, electronic control hydraulic knees, energy-storing feet, and shock absorbers.

DMD is an X-linked progressive neuromuscular disease characterized by muscle fiber degradation, motor deficits, and a shorter life expectancy. DMD is the most prevalent kind of muscular dystrophy, affecting around 0.01% in every 100 male births. The population of persons with DMD is predicted to expand as a result of technological improvements that have greatly enhanced their life expectancy. However, their independence-related quality of life remains low. By aiding with upper limb independence, robotic exoskeletons have the potential to improve their quality of life.

Immersive affective applications based on robotics and similar technology have received substantial consideration in the investigation and treatment of mental diseases. The immersion of emotional applications has been found to influence motivation and neurotherapy efficacy.

3.2 Robotic rehabilitation applications

Many neurological conditions including brain stroke, injury, and Parkinson's disorder might have disturbed the nervous system of a human being. These disrupted nerve signals in the brain cause improper movement of muscles, making it difficult to move ahead properly. Pace-training strengthens muscles and helps the brain to rebuild brain-neural trails. The popular robotic rehabilitation applications are as follows (Table 1.3):

Further, AMADEO, MYRO, PABLO, TYMO, and DriveSafety CDS 200 are other popular occupational therapy tools used in neurorehabilitation. Traditional surgery is a more invasive procedure with higher postoperative complications. AI-assisted surgery can reduce hospital stays and postsurgery infections.

Table 1.3 Popular AI robotic rehabilitation applications.

AI application	Major highlights
LokomatPro	**Highlights:** • Support treadmill training • Exclusively useful for severely impaired patients. • A harness safely supports the patient's body weight. **Tool type: Physical therapy**
Erigo Pro	**Highlights:** • Early rehabilitation incline table • Safe and early mobilization enabled • Supported walking activities **Tool type: Physical therapy**
Andago	**Highlights:** • Detects and actively follows a patient's movement using mobile robotic technology. • Exclusively useful for severely impaired patients. • A harness safely supports the patient's body weight. **Tool type: Physical therapy**
InMotion ARM	**Highlights:** • Offers evidence-based neurorehabilitation • The interactive technology of the robot constantly adopted patient's ability. **Tool type: Occupational therapy**
Diego	**Highlights:** • Robotic-assisted arm and shoulder rehabilitation • During therapy, it displays entertaining videos for further motivation and encouragement of patients. **Tool type: Occupational therapy**

A rehabilitation robot can relieve clinicians of time-consuming training tasks, analyze data collected by the robot during the training process, and assess the patient's rehabilitation progress [9]. Robotic-assisted machines, such as continuous passive motion (CPM), are effective in treating shoulder pain in stroke patients [10].

This type of rehabilitative robotics connects the human body in a wearable manner, controlling body movements during rehabilitation. The primary goal of rehabilitation training is to restore lower limb motor functions in patients. As a result, during rehabilitation training, a normal gait pattern is required as a reference input to the control system, a training goal, and a rehabilitation evaluation standard [11]. This rehabilitation robotics aids in reducing the therapist's workload in terms of energy efficiency during training, evaluation, and recovery assessment [12].

4. High-tech collaborations driving neural rehabilitation

4.1 Human–robot control interfaces

The study of interactions between humans and robots is known as human–robot interaction. The term "HRI" is frequently used by researchers. Human–computer interaction, artificial intelligence, robotics, natural language processing, design, and psychology all make contributions to the multidisciplinary field of human–robot interaction. With the development of AI, research is concentrating on one aspect of the safest physical interaction as well as on a socially acceptable interaction, based on cultural standards. The objective is to develop natural and intuitive speech, gesture, and facial expression interaction with the robot.

4.1.1 Digital–neural interfaces

Neuroelectronic interfaces, also known as neural or brain–computer interfaces, allow information to be transferred from the nervous system to an outside device. Typically, these devices are electrodes that record or modulate neuronal activity by delivering current to tissue (stimulation) or translating cellular activity into useful information (recording). Currently, neural interfaces are used in both research and clinical settings, providing answers to fundamental neuroscience questions about behavior, information encoding, and injury mechanisms, as well as providing direct control over prosthetic limbs, other peripheral devices, and hearing loss with cochlear implants and deep brain stimulation.

4.1.2 Electromyography

Neuroelectronic interfaces, also known as neural or brain–computer interfaces, allow information to be transferred from the nervous system to an outside device. Typically, these devices are electrodes that record or modulate neuronal activity by delivering current to tissue (stimulation) or translating cellular activity into useful information (recording). Currently, neural interfaces are used in both research and clinical settings, providing answers to fundamental neuroscience questions about behavior, information encoding, and injury mechanisms, as well as providing direct control over prosthetic limbs, other peripheral devices, and hearing loss with cochlear implants and deep brain stimulation.

- Tingling
- Numbness
- Muscle weakness
- Muscle pain or cramping
- Certain types of limb pain

EMG results are frequently required to assist in the diagnosis or exclusion of a number of conditions, including:

- Muscle conditions like polymyositis or muscular dystrophy
- Conditions like myasthenia gravis affect the connection between the nerve and the muscle
- Peripheral nerve disorders, such as carpal tunnel syndrome or peripheral neuropathies, affect the nerves outside the spinal cord.
- Conditions like polio or amyotrophic lateral sclerosis affect the motor neurons in the brain or spinal cord.
- Conditions that harm the nerve roots, like a spinal disc that has herniated

4.1.3 Brain–computer interfaces

An external device, most frequently a computer or robotic limb, and the electrical activity of the brain can directly communicate through a brain–computer interface (BCI), also known as a brain–machine interface (BMI) or smart brain. BCIs are frequently used to study, map, support, improve, or restore human cognitive or sensory-motor functions (Fig. 1.3) [1].

Based on how close electrodes are placed to brain tissue, BCI implementations range from noninvasive (EEG, MEG, EOG, MRI) to partially invasive (ECoG and endovascular) to invasive (microelectrode array) (Fig. 1.4) [2].

4.2 Neurorobotics

The combined study of neuroscience, robotics, and artificial intelligence is known as neurorobotics. It is the study of embodied autonomous neural systems in science and technology. Artificial spiking neural networks, large-scale simulations of neural microcircuits, and real biological systems are examples of computational models of neural networks inspired by the brain (such as connectionist networks) (e.g., in vivo and in vitro neural nets). Machines with mechanical or other physical actuation can incorporate such neural systems. This includes wearable technology, prosthetic devices, and robots. It also includes micromachines on a smaller scale and furniture and infrastructure on a larger scale.

The field of neuroscience that combines robotics and deals with the study and application of science and technology of embodied autonomous neural systems, such as brain-inspired algorithms, is known as neurorobotics. It is predicated on the notion that the body is embedded in the environment and that the brain is

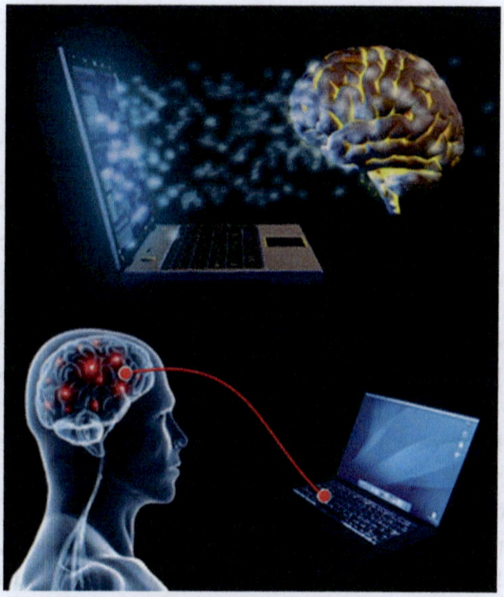

FIGURE 1.3

Brain–computer interface with neuro-robotics.

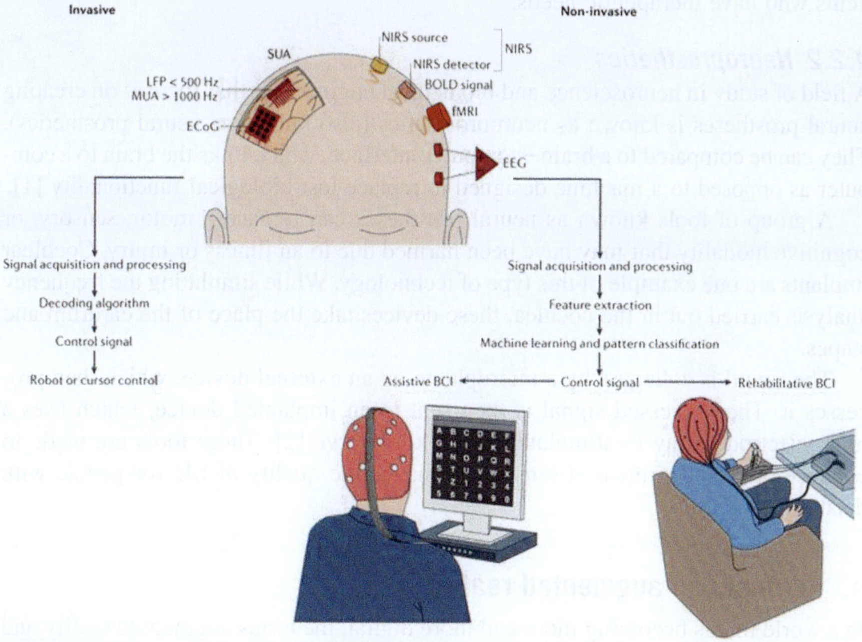

FIGURE 1.4

Brain–computer interfaces.

embodied. As a result, rather than operating in a virtual environment, the majority of neurorobotics need to operate in the real world.

4.2.1 Neuroexoskeletons

Exoskeletons are wearable devices that support and facilitate movement or increase the body's capacity. Exoskeletons are tools that help or improve a person's physical capabilities. Exoskeletons were created with the idea of giving the user external support so they could carry out a specific task. Exoskeletons are constructed from a network of electro-mechanical devices that attach to the human body's extremities.

Clinical trials conducted by Esko have included, among other things, the management of spinal cord injuries, neurogenic bowel, robotic rehabilitation, and stroke rehabilitation. The EksoNR provides support for the postural trunk, knee, hip, and ankle. By detecting patients' instability movements and providing feedback to support them while walking or standing, this device encourages lower extremity movement in the patient.

It has been demonstrated that the upgraded version of Ekso GT offers greater advantages for neurorehabilitation. Data capture, posture support, clinician control, and smart assist software are just a few of the features offered by the technology. The Food and Drug Administration (FDA) approved this device as the first of its kind for therapeutic purposes for stroke, acquired brain injury (ABI), and now MS. With these indications, the company hopes to improve the health of more patients who have therapeutic needs.

4.2.2 Neuroprosthetics

A field of study in neuroscience and biomedical engineering that focuses on creating neural prostheses is known as neuroprosthetics (also known as neural prosthetics). They can be compared to a brain–computer interface, which links the brain to a computer as opposed to a machine designed to replace lost biological functionality [1].

A group of tools known as neural prostheses can replace a motor, sensory, or cognitive modality that may have been harmed due to an illness or injury. Cochlear implants are one example of this type of technology. While simulating the frequency analysis carried out in the cochlea, these devices take the place of the eardrum and stapes.

The sound is collected by a microphone on an external device, which then processes it. The processed signal is then sent to an implanted device, which uses a microelectrode array to stimulate the auditory nerve [2]. These tools are made to enhance or replace impaired senses to enhance the quality of life for people with disabilities.

4.3 Virtual and augmented reality

In a world that is becoming more and more digital, the terms augmented reality and virtual reality are crucial. Despite being two distinct technologies, both terms are frequently used interchangeably. But how do AR and VR compare and contrast with one another?

FIGURE 1.5

Virtual and augmented reality.

In augmented reality (AR), real and virtual elements are combined. It is a technology that works well on both desktops and mobile devices. The ability to reflect digital elements in the physical world is what distinguishes it as unique (Fig. 1.5).

The main distinction between AR and VR is that the latter uses computer simulations. This indicates that a graphic representation of reality or a fictional world is created.

The user can fully immerse themselves in the digital world by using the right hardware. As a result, there are significant differences between VR and AR headsets. Sensory components that convert physical movements into a modeled reality are needed for hardware designed for virtual reality (Fig. 1.6).

FIGURE 1.6

Physical movements in a modeled reality.

There are very substantial differences exists between augmented reality and virtual reality. But this obvious distinction does not imply that one of the two technologies is superior to the other. Rather, both technologies excel in distinct application domains:

- AR enhances a real-world scene, while VR creates an immersive virtual environment.
- Compared to AR, which is only 25% virtual, VR is 75% virtual.
- Unlike AR, VR requires a headset device.
- AR users interact with the real world, whereas VR users move in a wholly fictional environment.
- Compared to VR, AR requires more bandwidth.

The goal of augmented reality is to improve both the real and virtual worlds. Virtual reality (VR) replaces the real world with a made-up reality that is primarily used to improve video games. A symbiotic relationship of superior systems is created when AR and VR are used together. Even though they can function separately, together they provide users with a better and more interesting experience. The idea behind this is to construct a world of fiction that permits communication with the real world.

4.4 AI algorithms

- Intelligent systems can be created using a variety of techniques, including machine learning, probabilistic reasoning, search and optimization, logic, and planning. The majority of traditional AI methods were not intended to function in dynamic, unpredictable, and nondeterministic environments.
- The most successful robot applications over the past few decades have been limited to straightforward tasks involving known circumstances; because it is frequently used in highly repetitive, rarely varied, and dexterity-required processes like those carried out in industrial plants, robot automation has experienced tremendous commercial success. Physical interactions and human communication, however, understanding to deal with a wide range of complex semantics.
- Monte Carlo localization (MCL) is the most popular technique for locating the robot in its surroundings. Based on prior knowledge, the particle distribution is initially uniform. A recursive Bayesian estimation is used to resample the particles.
- This cycle is repeated until the robot's actual position is reached by all the particles. Sometimes there is not a map of the surroundings. In these scenarios, the robot needs to gather a map as it moves through the environment. Simultaneous localization and mapping (SLAM) is a name for this issue. Usually, probabilistic methods like the extended Kalman filter are used to solve this issue.

- Different tasks typically call for distinct features and particular methods to extract them. There are numerous feature extraction techniques available in the field of computer vision research, including those for color, texture, shape, and motion. To identify objects with distinct color components, color can be used.
- The deep-learning algorithm, which is also known as the hierarchical-learning algorithm, makes an effort to learn at various levels, each of which corresponds to a different level of abstraction. Big data, also known as extremely large datasets, are difficult to process, manage, and maintain but are necessary for deep learning. Additionally, existing deep-learning models need to be labeled in the form of training data. Artificial neural networks (ANNs) are frequently used in the deep learning era. Deep neural networks (DNNs) process features using several layers, and typically each layer extracts some data. Higher-level features in this architecture are represented in terms of low-level semantics.
- Deep learning enables computational models made up of numerous processing layers to learn data representations with varying degrees of abstraction. These technologies have significantly advanced the state of the art in voice recognition, visual object identification, object detection, and a variety of other fields such as drug development and genomics [14]. They use multimodal and high-dimensional sensor data that are typically initiated in robotic systems to create compact representations of states.
- Ambient intelligence (AMI) is a type of AI that includes environments that are aware of and respond to human presence. This system is integrated into a patient's home and their priorities. This analysis is the individual's real time in terms of his/her geographical location within the given environment, including the time, as well as his/her physical environment, including home/office or rehabilitation setting, capturing the individual's needs and function.

5. Selection of AI-based supportive technologies and tools for instigating neurorehabilitation

Neurological rehabilitation aims to augment current functionality or restore the functionality of damaged neurological circuits, to the extent that this is possible. A deeper comprehension of the neurophysiological details underlying poor motor control should be given priority to increasing the clinical impact of technological advancements in rehabilitation [13]. To select AI-based supportive technologies and tools for instigating neurorehabilitation, the following criteria must be adopted by clinical practitioners.

- AI-based aids for corrective and accurate analysis of medical imaging
- To reduce the cost of developing new pharmaceuticals.
- For unstructured data analyses.

- To create sophisticated and integrated drug discovery platforms in a short time in an economic way.
- For emergency medical personnel.
- To study and treatment of cancer, particularly in radiation therapy
- Data collected for predictive analytics.
- Genetic medicine's discovery and advancement accelerated.
- To produce the best health outcomes, the AI and ML industries must create healthcare systems and tools that guarantee fairness and equality in data science and clinical studies.

6. Conclusions and future directions

This chapter has a review of the most important robotic technologies for neural rehabilitation, a discussion of the challenges they face in development and use, as well as a look at new directions the field is taking. Despite having various unfolded issues, these technologies resolve unconditionally clinical problems and yield collaborative benefits in the era of neurorehabilitation. Better clinical practice and patient care may result from the collaboration of human intelligence and intelligent algorithms. Additionally, humans are needed to assess machine recommendations, delegate authority to them, and exercise control over them. Therefore, since the use of AI-based technology is likely to increase soon, learning to implement AI in rehabilitation is very important for quality services. The ability to connect, learn successfully over the course of our lives, and come up with innovative solutions to problems that matter to us will be essential in an era of intelligence.

References

[1] Nayak S, Das RK. Application of artificial intelligence (AI) in prosthetic and orthotic rehabilitation. Service Robotics; 2020.

[2] GBD 2016 Neurology Collaborators. Global, regional, and national burden of neurological disorders, 1990–2016: a systematic analysis for the Global Burden of Disease Study 2016. The Lancet Neurology 2019;18(5):459–80.

[3] Chandra J, Rangaswamy M, Banerjee B, Prajapati A, Akhtar Z, Sakauye K, et al. Applications of artificial intelligence to neurological disorders: current technologies and open problems. In: Augmenting neurological disorder prediction and rehabilitation using artificial intelligence. Elsevier; 2022. p. 243–72.

[4] Zhang Y, Yu H, Dong R, Ji X, Li F. Application prospect of artificial intelligence in rehabilitation and management of myasthenia gravis. BioMed Research International 2021;2021:5592472.

[5] Zhiguo W, Baofeng Y. Polypharmacology. Springer Science and Business Media; 2022.

[6] Ghazal S, Sauthier M, Brossier D, Bouachir W, Jouvet PA, Noumeir R. Using machine learning models to predict oxygen saturation following ventilator support adjustment in critically ill children: a single center pilot study. PLoS One 2019;14(2):e0198921.

[7] Acampora G, Cook DJ, Rashidi P, Vasilakos AV. A survey on ambient intelligence in health care. Proceedings of the IEEE. Institute of Electrical and Electronics Engineers 2013;101(12):2470–94.
[8] Nizamis K, Athanasiou A, Almpani S, Dimitrousis C, Astaras A. Converging robotic technologies in targeted neural rehabilitation: a review of emerging solutions and challenges. Sensors 2021;21:2084.
[9] Rabhi Y, Mrabet M, Fnaiech F. Intelligent control wheelchair using a new visual Joystick. Journal of Healthcare Engineering 2018:1–20.
[10] Beasley RA. Medical robots: current systems and research directions. Journal of Robotics 2012;20:1–14.
[11] Vélez-Guerrero MA, Callejas-Cuervo M, Mazzoleni S. Artificial intelligence-based wearable robotic exoskeletons for upper limb rehabilitation: a review. Sensors 2021;21(6):2146.
[12] Backus D, Winchester P, Tefertiller C. Translating research into clinical practice: integrating robotics into neurorehabilitation for stroke survivors. Top Stroke Rehabilitation 2010;17:362–70.
[13] Russell S, Norvig P. Artificial intelligence: a modern approach. Upper Saddle River, NJ: Prentice Hall; 2009.
[14] Hillman M. Rehabilitation robotics from past to present–a historical perspective. In: Advances in rehabilitation robotics. 1st ed. Springer; 2004. p. 25–44.

Further reading

[1] Sharma K, Manchikanti P. Regulation of artificial intelligence in drug discovery and health care. Biotechnology Law Report 2020;39:371–80.
[2] Paul M, Mishra P, Reddy KG, Pulugurtha A. Evaluating degradation resistance of silicone rubber nanocomposites using deep learning technique for outdoor high voltage applications. 2nd International Conference on Intelligent Technologies. 2022.
[3] Aymé S, Bellet B, Rath A. Rare diseases in ICD11: making rare diseases visible in health information systems through appropriate coding. Orphanet Journal of Rare Diseases 2015;10:35.
[4] Arora A, Gupta R. A comparative study on application of artificial intelligence for quality assurance in manufacturing. 4th International Conference on Inventive Research in Computing Applications (ICIRCA). 2022. p. 1200–6.
[5] LeCun Y, Bengio Y, Hinton G. Deep learning. Nature 2015;521(7553):436–44.
[6] Kim MS, Kim SH, Noh SE, Bang HJ, Lee KM. Robotic-assisted shoulder rehabilitation therapy effectively improved poststroke hemiplegic shoulder pain: a randomized controlled trial. Archives of Physical Medicine and Rehabilitation 2019;100(6):1015–22.

CHAPTER 2

Complex approaches for gait assessment in neurorehabilitation

Ligia Rusu and Mihnea Ion Marin
University of Craiova, Craiova, Romania

1. Introduction

Neurorehabilitation is a very complex and dynamic process that is based on neuroplasticity. Neuroplasticity is one of the most important nervous system properties. The aim of neurorehabilitation is to limit the effects of neuromotor impairments and improve the quality of life. In the present scenario, a holistic approach is mandatory for patients who undergo neurorehabilitation. Gait rehabilitation is one of the most important goals of rehabilitation. In this context, biomechanic evaluation of gait has to be carried out with new technologies. These technologies are used for designing neurorehabilitation strategies based on functional status. They are also used for monitoring the complete process of therapy. Adaptation to daily life with gait rehabilitation is one of the major objectives of this study. It is also a challenge for the neurorehabilitation process because it needs a complex evaluation process to identify the main problems of gait disorders. The difficulty in identifying the causes is mainly due to the lack of motor control. The choice of methods in gait assessment depends on etiopathogeny, level of injury, severity of nervous injury, clinical disorders, and rehabilitation capacity.

In this chapter, we present some theoretical aspects of gait analysis and present three study cases: hemiparesis in cerebral palsy (CP), paraparesis in postchemotherapy, and multiple sclerosis. The main aim of this study is to develop the link between gait analysis and clinical application. Kinematic analysis, kinetic analysis, and balance assessment help to know the neural control strategies and understand the abnormal pattern of gait. Kinematic analysis is useful to quantify the gait pattern. The kinetic analysis provides information about gait disorders and underlines the aspects of muscle function. Generally, gait refers to the movements associated with hip flexion, knee flexion, and ankle dorsiflexion. These provide the orthostatic position and dynamic balance. Hence, analysis of these movements during gait facilitates the neurorehabilitation process, which focuses on postural control and dynamic balance. The subjects of this chapter are people with CP and other neurologic pathologies. This gait analysis-based study provides information about therapeutic strategies. These people exhibit a lot of adaptive postural mechanisms.

Computational Intelligence and Deep Learning Methods for Neuro-rehabilitation Applications
https://doi.org/10.1016/B978-0-443-13772-3.00001-7

2. Theoretical approach of gait analysis

Gait analysis can be realized as the link between patient movement, patient capacity, and performance (means what a patient can do in fixed conditions and what he/her can do daily) [1]. However, this link is not yet properly established [2]. In the context of functional capacity, gait refers to the process of performing short tasks (assessed by the Gross Motor Function Measure (GMFM)) [3], level of endurance (assessed by the 1- or 6-Minute-Walk Test (1 MWT or 6 MWT)) [4,5].

Performance is most often assessed by self-perception and is based on a questionnaire about daily life [6] and physical activity habits and activity limitations [7,8]. Gait analysis is important in clinical rehabilitation, and it is used in a variety of applications. In neurorehabilitation, the problems of gait are linked with balance because of disorders of body systems. These disorders are involved with movement in various parts of the nervous system, muscle system, and sensorial systems. The objective of gait analysis is to identify different and wrong gait patterns that need treatment. Gait analysis must be correlated with posture disorders and include measures of step length, regularity of steps, speed, and rhythm.

In the clinical approach, the main role is to develop a functional diagnostic such as:

- Identify the sources of pain, segmentary impairments generated by the nervous central system.
- Generate a report in real time, which allows the clinician and physiotherapist to monitor the evaluation and create feedback
- Allow to create a personalized program

Koleva et al. [9] talk about functional rehabilitation and about the requirement for complex evaluation due to somatosensorial disorders, cognitive disorders, and lack of motor control. All these are based on neuroplasticity.

Measure of kinetic and kinematic joints is the most important approach in gait analysis for setting the neurorehabilitation. These are complementary to electromyographic (EMG) assessment, energy consumption (VO_2), and plantar pressure. Brand [10] provides information on the need for gait analysis in neurorehabilitation, setting the functional diagnostic, setting the level of severity of ethiopatogeny, mechanism of impairments, and setting the therapeutic plan.

3. Gait pattern—theoretical and practical framing

Study of human movement and gait involves a lot of tools such as scanners, sensors, and video systems. The results are the gait patterns that allow to have a feature of the clinical aspects. We can have these patterns to identify the gait parameters for lower limb joints, which can be compared with the reference patterns proposed by Perry [11]. These patterns help clinicians and physiotherapists to decide without a formal gait evaluation [12]. Many clinicians associate the abnormal patterns of a group of patients with segmentary impairments [13–15]. The problem is how the gait analysis is made because are a lot of ways, a lot of questions if are or not specific gait pattern for different pathologies. Gait analysis is based on the biomechanics principle and evaluation of intersegmental joints. So, Baker [16] demonstrates that the Eular angle approach used for estimating the pelvic angle can provide wrong conclusions from the clinical point of view. He proposes an alternative way, and this will be adopted at a large level. The method for interpreting the tridimensional angles, more precisely rotations, is not clearly understood by clinicians and bioengineering specialists [17]. Baker attempts to make a standardization of the process of reporting the angle but the debate continues [18,19].

Almost all commercial equipment used for gait analysis employ the conventional Gait Model [20]. They are referred to as the Newington, Gage, Davis [21], Helen Hayes, Kadaba [22,23], or Vicon Clinical Manager (VCM) model. This was developed using the minimum number of markers to determine three-dimensional kinematics and kinetics [24,25]. This model assumes three free grades for hip and knee joints and two free grades for ankle joints. The model is hierarchical and requires the detection of the proximal segments. Hence, the distal segments can be defined and incorporated into the regression equations to determine the position of the hip joint center with respect to pelvic markers.

The joint moments are presented with orthogonal system axis, which is fixed on the distal or proximal side, or sometimes toward the reference system of the laboratory [26,27]. Reporting the joints moments to the orthogonal system of the axis and reporting the joint rotations to nonorthogonal system led to difficulties in understanding the change in the joint moments. Some authors have given some explanation but it is not clear to interpret. Probably, the most important aspect that limits the understanding in the present scenario is that this analysis is descriptive.

Many authors from the scientific community consider that gait analysis, from the kinematic point of view, is a descriptive analysis but they confirm that kinetic analysis can explain gait patterns. This approach seems to be wrong. The kinetic analysis includes a lot of measurements. Recent papers discuss the correlation between the results of these measurements and the accuracy of the experimental data [28,29]. Gage [11,30] and some studies analyze the results by comparing them with Perry's results [31], which allow to conclusions about the impairments. The lack of many studies is that a small number of studies give explanations about how to make the difference between normal and pathological gait.

4. Clinical applications of gait analysis in neurorehabilitation

Gait analysis (GA) is a research domain that has applicability in medical science, sport science, and in the evaluation of prosthesis or surgical interventions [32,33], therapeutic plan evaluation of fall risk [34], and evaluation of neuropathy [35].

All these aspects need to be taken into consideration for the neurorehabilitation program. Kinematic assessment of gait estimates the expected outcomes regarding gait rehabilitation and the future conditions regarding functionality [36] and monitoring the intervention [37–39]. Usually in clinical activity, the gait analysis is made using standardized scales and tests [40]. These tools are easy to use as they do not need much time. However, the specificity and accuracy are very less [41]. Hence, new technologies must be used. Usually, gait analysis refers to hip, knee, and ankle analysis. However, pelvic analysis is also important because it is often a constraint for movement [42].

Rudroff and Proessl inform that even if many studies talk about significant muscle asymmetry, postural stability and gait are the results of many factors [43]. One of the ways for analyzing the asymmetry between the left and right lower limb or inside of the lower limb (intersegmental assessment) is the diagram, called cyclogram. The cyclogram in gait analysis represents the closed trajectories. It reflects the kinematic of gait during the gait cycle [43]. For example, cyclograms of the hip–knee are the results of drawing the hip angle with respect to the knee joint. It shows us the relationship between these two joints. In other words, cyclograms offer information about interjoints coordination. However, interpreting cyclograms is not easy [44–46].

In the following section, we present three pathologies: multiple sclerosis, cerebral palsy, and paraparesis postchemotherapy.

4.1 Multiple sclerosis

Multiple sclerosis (MS) is a demyelinating disease of the central nervous system (CNS), which is progressive in nature. It is one of the main clinical features of gait disorder. It is mainly due to the disturbance in the process of nervous conduction and lack of motor control. MS is studied by different kinematic analyses because muscle fatigue contributes to the development of gait disorders. Analysis of the effect of fatigue could be realized by biomechanic evaluation of the lower limb [47]. Despite the research works on gait analysis in MS, only a small number of studies analyze the kinematic asymmetry of the lower limb during gait. The main problems of coordination and motor control are developed at hip, knee, and ankle. This asymmetry is more obvious in mild MS at the ankle level.

The existence of a mild correlation between symmetry parameters and gait parameters suggests that symmetry parameters influence gait quality and gait efficiency. It is mainly due to the reason that MS patients have a great asymmetry, reduction of speed, and step length during the double support gait phase. Many research works suggest that it is possible to have a differentiated change of functionality of the lower limb in MS patients. However, some studies consider that the main way to analyze this aspect is by evaluating the asymmetry of spatiotemporal gait parameters. This could reflect the kinematic muscle and joint asymmetry. In this context, biomechanic parameters of gait and the creation of the cyclogram assist to assess the right and left symmetry and interjoint symmetry. This assists in assessing the disorders of motor control in MS [48]. There is a need for more research to identify the link between gait symmetry and MS evolution.

Kinematic changes of gait in MS, reduction in knee flexion during the swing phase of gait, and reduction of hip extension during the terminal phase and preswing phases are present in MS patients with lower levels of clinical impairments [49,50]. The main features of gait parameters in MS are as follows: speed, endurance, step length, cadence, joint range of motion (ROM), metabolic cost, and quality of gait. A standard set of gait parameters is possible only with observational evaluation and 3D analysis of the gait.

4.2 Hemiparesis

Hemiparesis is one of the common impairments in stroke or CP that affects the gait. Sometimes, the patient could be walking but cannot perform the daily activities [51]. Many studies talk about the main and important problems in hemiparesis. The main problem is ankle movement, which is disturbed by triceps sural spasticity [52]. In Europe, the gait after stroke or in CP was assessed by clinical evaluation and observation of the gait cycles. This approach is in accordance with the somatosensorial system function [53]. Gait capacity is one of the most important factors that influence the severity of disability [54]. In these situations, speed gait and cadence decrease, gait duration increases, and the paretic lower limb has a long swing phase and a short support phase [55].

A detailed study of the literature provides information about the gait features in hemiparesis [56]. It also shows a decrease of knee and hip ROM in the sagittal plane [43]. At the same time, gait perimeter and area of balance or total area are the parameters that could be taken for consideration in assessing the interjoints coordination by cyclogram of hip–knee couple or knee–ankle couple [57,58]. It is necessary to analyze each gait phase [59]. Knee joint hyperextension of the knee and decrease of knee flexion during the swing phase are the features of stroke due to extensor knee spasticity and proprioceptive disorders [60,61].

An important aspect is the reduction of area for the stance phase, which may be generated by the disorders of hip and knee joints. This parameter could be a significant parameter from the clinical point of view to classify stroke patients. If we speak about CP, the main problem from the neurorehabilitation point of view is the ankle joint, which has no motor control, which is due to a lack of motor learning [62].

Zhou [63] describes six gait strategies with respect to the mass center movements such as:

- Pelvis tilt
- Pelvis rotation
- Ankle and knee movements
- Lateral movement of pelvis

Anca Irina Grigoriu et al. [64] compare the temporal-spatial gait parameters of children affected by CP and children with normal development. The authors also observed that the CP children have bilateral gait parameters and pelvis asymmetry. Another similar study is carried out by Ragab K. Elnaggarab [65]. This observational and descriptive study includes 42 children with CP (age 5–8 years old). The author makes a gait analysis of kinematic parameters such as pelvis rotation, hip and knee ROM, and progressive angle of the foot. An analysis on temporal-spatial gait parameters such as speed and step length was also carried out. Here, ankle rotation seems to influence the single support phase. The author observes that the speed of gait and symmetry are affected by abnormal hip rotation. Gait symmetry is influenced by deviation of foot axis and ankle rotation.

4.3 Paraparesis postchemotherapy (postchemotherapy polyneuropathy CIPN)

This pathology is more commonly seen in the last years and needs a complex approach because they develop a lot of dysfunctionalities. Tofthagan, Overcash, and Kip [66] observe that postchemotherapy increases the risk of fall two times more than the people who do not receive chemotherapy. Also, they observe a decrease in tactile and proprioceptive sensibility, a decrease in muscle force, which leads to gait disorders, balance disorders, and inferior neuromuscular control. All these are serious problems at the moment in every chemotherapy dose administration. The main form of evaluation in this type of pathology are clinical scales like Berg balance, Timed Up and Go, and Single Leg stance tests to verify levels of balance dysfunction caused by CIPN. Biomechanic assessment of gait in CIPN is rarely seen in the literature.

5. Case studies

In the following section, we present three case studies. Kinematic analysis of gait parameters is done with the VICON system from Bioengineering Laboratory (www.incesa.ro). The system includes 14 infrared cameras with a frequency of 200 Hz. The system uses NEXUS software and a special skeleton model named Plug-in Gait for recording the movements associated with each subject. The subjects wear a black suit that has 39 reflective markers with a diameter of 14 mm. It is shown in Fig. 2.1. Marker positions are in accordance with the protocol described by Davis et al. [67].

Using the software, we made the movement analysis of the lower limb during the gait on a plane surface at a comfortable speed. It follows the trajectories of the main important points on the lower limb (knee and malleolus) and also a variation of angular amplitude for the hip, knee, and ankle joints. Participants walked at a self-selected speed in the most natural manner possible on a 10 m walkway at least three times, with suitable rest times.

The values of flexion–extension angles are compared with reference values proposed by Perry in his protocol [68].

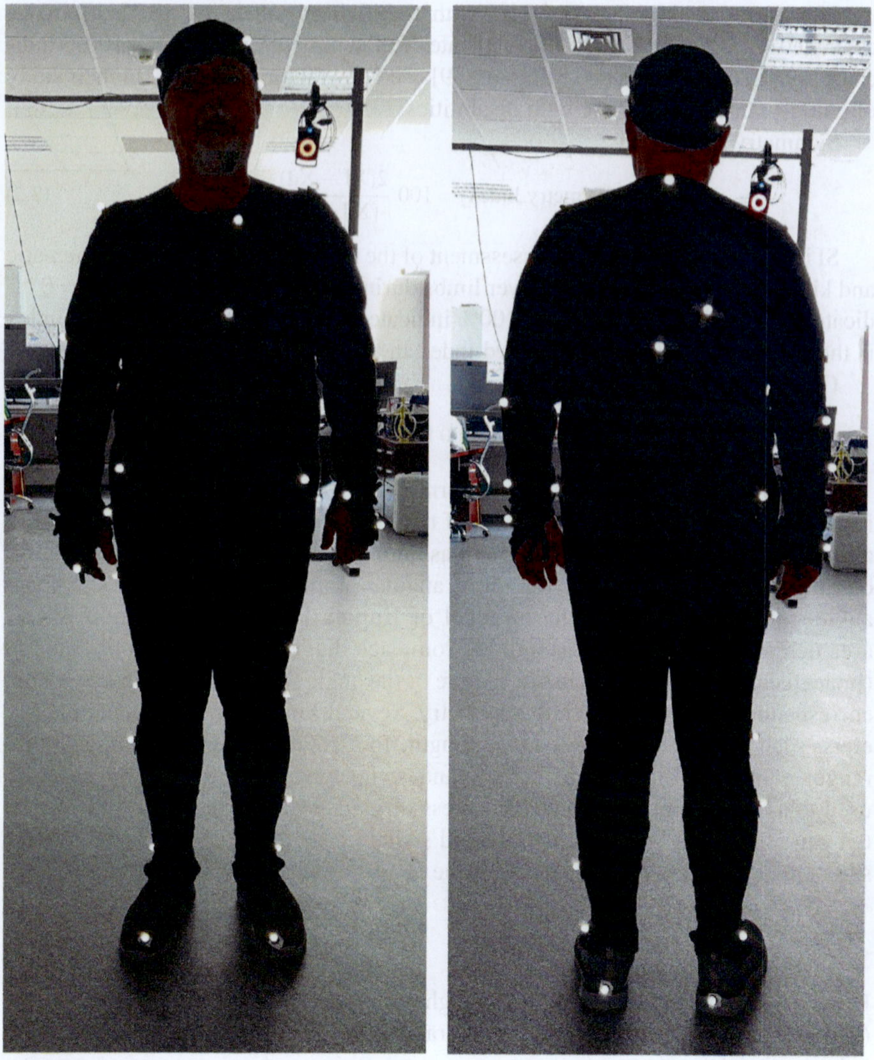

FIGURE 2.1
Vicon marker setup.

In our study, the recorded values are used for the evaluation of the symmetry of the kinematic parameters of the lower limb. The symmetry has been demonstrated by qualitative shapes using cyclograms. The following indexes are used to analyze the cyclograms.

Symmetry ratio (R):

$$R = \frac{Xs}{Xd} \tag{2.1}$$

This factor decides the variables with the highest value, and as such creates asymmetries. The value of RI = 0 indicates full symmetry, while RI $\geq$ 100% indicates asymmetry. Andres and Stimmel [69] used this factor effectively in their study of gait asymmetry in people with disabilities.

Symmetry index (SI):

$$\text{SI (Symetry Index)} = 100\,\frac{2(\text{Xl} - \text{Xd})}{(\text{Xd} + \text{xl})}\,[\%] \tag{2.2}$$

SI is a process of percentage assessment of the differences between the kinematic and kinetic parameters for both lower limbs during walking. The value of SI = 0 indicates full symmetry, while SI $\geq$ 100% indicates its asymmetry [70]. The SI index is the most commonly used and cited index in publications on gait symmetry.

Gait asymmetry (GA):

$$\text{GA} = 100\,\ln\frac{\text{Xs}}{\text{Xd}}\,[\%] \tag{2.3}$$

This equation is a logarithmic transform of the RI factor. Plotnik et al. [71] used it to calculate asymmetry on the basis of the duration of the swing phase. GA = 0 and GA $\geq$ 100% denote symmetry and asymmetry, respectively. Based on the recordings, we observe the cyclogram of angular amplitude for the hip, knee, and ankle–foot complex using the protocol of Kutilek [72]. The cyclogram has an area delimited by the curve generated from each diagram. Cyclogram orientation (in degrees) is given by the absolute value of the angle formed by the 45 line which corresponds to perfect interlimb symmetry. Several kinematic and kinetic parameters, such as walking speed, stride length, foot rotation angle, maximum joint range of motion, duration of stance, and swing phases of gait, are frequently used to assess and monitor the correctness of gait. We underline that we describe the gait phases like the stance phase and swing phase from the gait cycle. All results are analyzed in accordance with the results of Perry's gait values.

5.1 Case study 1

Patient 1 (P1), female, 55 years old, weight 60 kg, height 1.76 m, body index mass (BMI) 20.76, paraparesis postchemotherapy. Clinical evaluation—gait disorders, balance disorders.

Kinematic parameters—The results regarding kinetic parameters (range of motion) are presented in Table 2.1 and the average values of indices (RI, SI, GA) are given in Table 2.2.

According to reference values for ROM, we observe that the ROM is only 63.85% from normal ROM for hip. The ROM is 76.18% from normal ROM for knee, and the ROM is 80.96% from normal ROM for ankle. From these results, we can conclude that the main problem is at the hip joint level due to the limit of extension.

A gait asymmetry, generated by step length and step high, is observed. We can observe an asymmetry in SI, which means that the kinetic and kinematic parameters have symmetry, but no gait symmetry. In connection with the behavior of the lower

Table 2.1 Average values of range of motion (hip, knee, and ankle joints—sagittal plane).

	ROM (knee)	ROM (ankle)	ROM (hip)
P1	45.71 degrees	20.24 degrees	25.51 degrees

Table 2.2 Values of kinematic parameters and symmetry indexes.

P1	Right	Left	RI	SI	GA
Step high (m)	0.128	0.142	−10.938	0.104	10.380
Step length (m)	0.635	0.717	−12.913	0.121	12.145
Speed gait (m/s)	0.49	0.53	−8.163	0.078	7.847
Cadence (steps/min)	46.04	44.44	3.475	0.035	−3.537

limb joint during the gait cycle, we make an analysis of each joint, during a gait cycle, and present the evaluation for the main gait phases. They are shown in Figs. 2.2–2.5. Table 2.3 shows the information about stance phase and swing phases.

We observe that patient P1 presents a decrease in swing phase duration and an increase in stance phase, more for the right lower limb. The evaluation of ROM

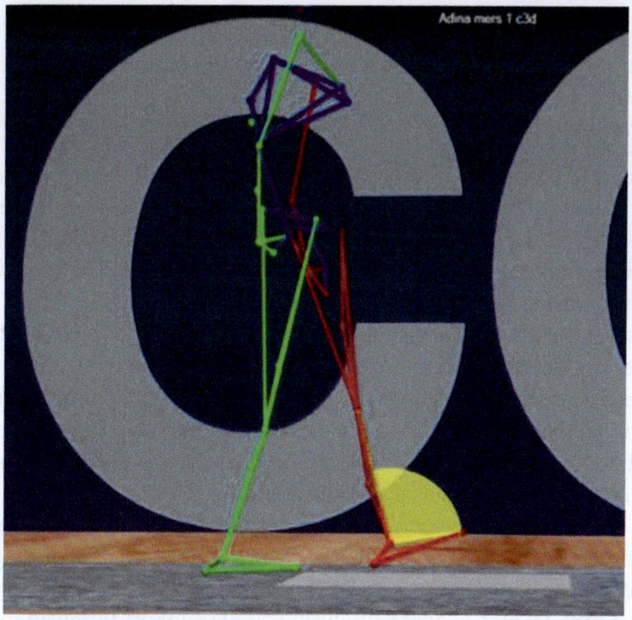

FIGURE 2.2

Heel strike (initial contact).

FIGURE 2.3

Stance phase.

for each joint is shown in Figs. 2.6–2.8. The results are presented in comparison with the reference values model proposed by Perry.

Interpretation of the results:

Hip joints—we observe a limitation of mobility, lock of the hip on flexion at 10 degrees in single stance support in the right hip. These are associated with an increased stiffness of the hip joint and lack of hip extension. During the swing phase, we observe a high level of flexion hip movement. The most important key for propulsion in the lower limb is trunk movement. For the left hip, the stiffness is a little bit lesser than the right side. The flexion movement is around 30 degrees in double support 1 (initial) and single support. A hip extension of 12 degrees is observed during the stance phase followed by limit flexion during the swing phase. The main aspect is the limitation of hip joint extension and block of both joints. The main key on propulsion seems to be trunk movement.

Knee joints—Right joint has a flexion of around 15 degrees during single support and on double support 2 (preswing). The knee flexion is small at the beginning of the swing phase, which indicates a high stiffness of the knee. The knee movement is normal on the left knee during double support 1. An extension of 8 degrees is observed during single support followed by a flexion of 49 degrees.

FIGURE 2.4

Preswing (toe off).

Ankle joints—A high stiffness during single support is observed on the right side. The ankle is fixed in dorsiflexion at an angle of 10 degrees. This is correlated with knee position, which is blocked in flexion. During the stance phase, the foot has stiffness with plantar flexion of 5 degrees. We can observe that during double support 1, the foot is in dorsiflexion at 12 degrees for the left side. It is 0 degree during single support and double support 2. At this moment, the ankle movement has a sudden plantar flexion, During the swing phase, the foot has a high stiffness and it is locked in dorsiflexion at −5 degrees, which is −18 degrees lesser than Perry value. If we compare right and left sides, we can conclude that the left hip has a small flexibility than the right side. The right foot has the plantar flexion position during the gait cycle. Cyclograms for two couples such as knee–hip and knee–ankle foot are useful for the analysis of the gait. The cyclograms are presented in Figs. 2.9–2.11.

In the first cyclogram, we observe a lot of jerky movements which is due to the lack of motor control of the ankle joint during the stance phase at the border between double support 1 and single support.

In Fig. 2.10, we can see the relationship between knee and hip. We observe that the hip extension does not exist in double support 2. In the single support phase, a lot of oscillations are observed at the knee and hip joints along with jerky movements.

FIGURE 2.5

Swing phase.

Table 2.3 Components of gait cycle phases.

	Stance phase			Swing phase
	Double support (% from gait cycle)	Single support (% from gait cycle)	Double support (% from gait cycle)	Single support (% from gait cycle)
P1 right limb	24	27	29	20
P1 left limb	28	19	26	27

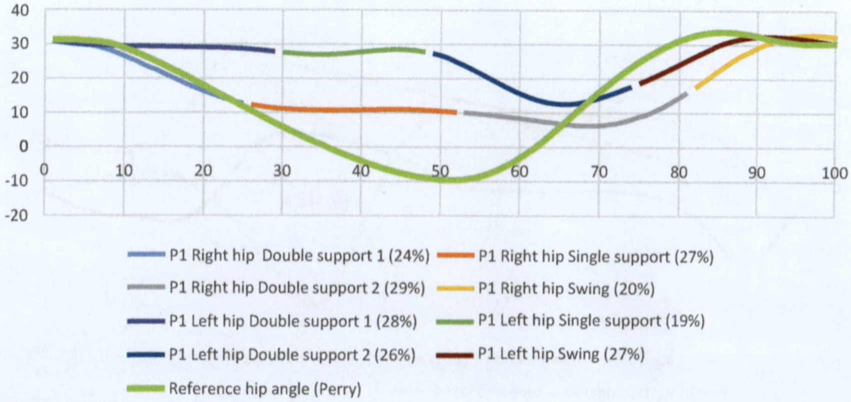

FIGURE 2.6

ROM for hip joints—right and left lower limbs.

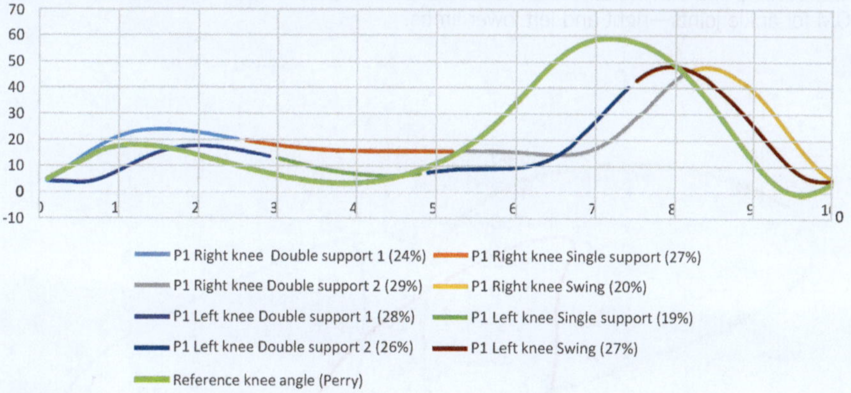

FIGURE 2.7

ROM for knee joints—right and left lower limbs.

At the same time, we observe a short time for the swing phase, which means a reduction capacity for maintaining the motor control of the lower limb. The cyclograms give us the possibility to analyze the symmetry for these two couples and this is presented in Fig. 2.11. It shows a good symmetry for knee joints than the other two joints such as hip and ankle.

FIGURE 2.8

ROM for ankle joints—right and left lower limbs.

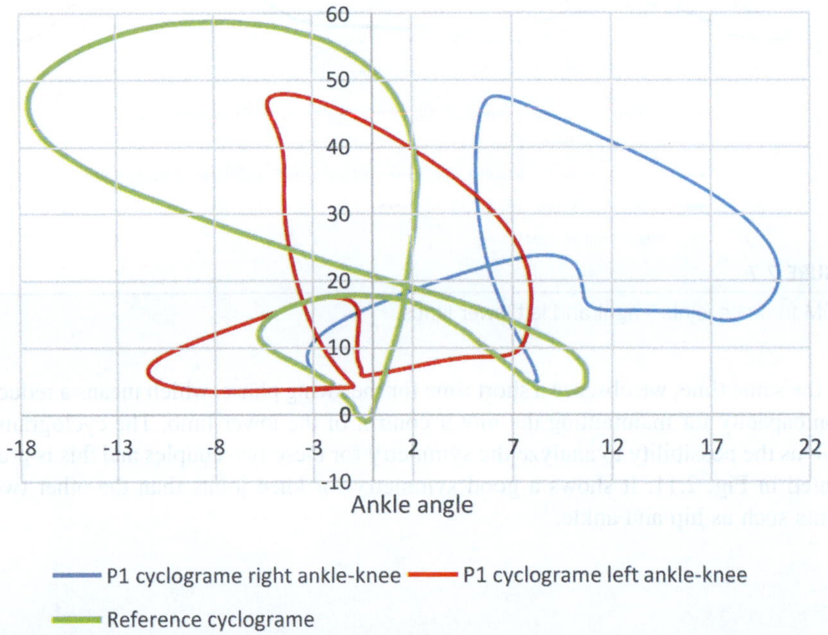

FIGURE 2.9

Cyclogram knee–ankle foot/gait phases.

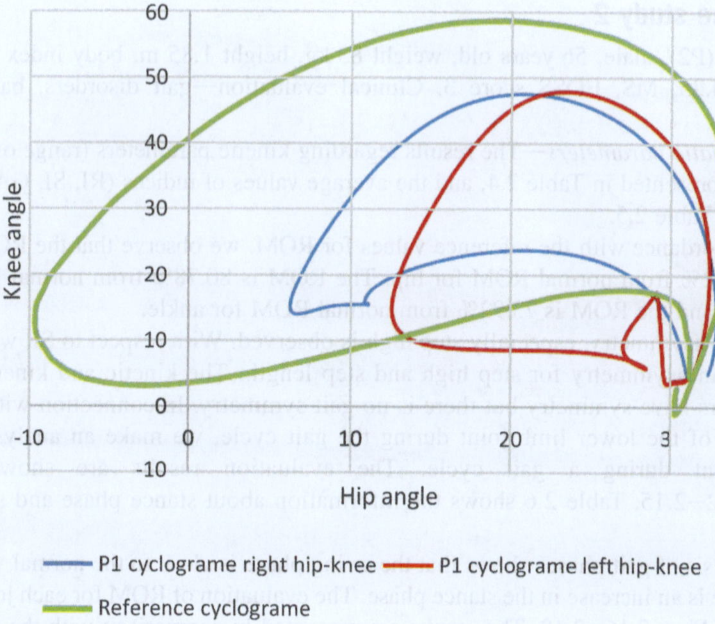

FIGURE 2.10

Cyclogram knee–hip/gait phase.

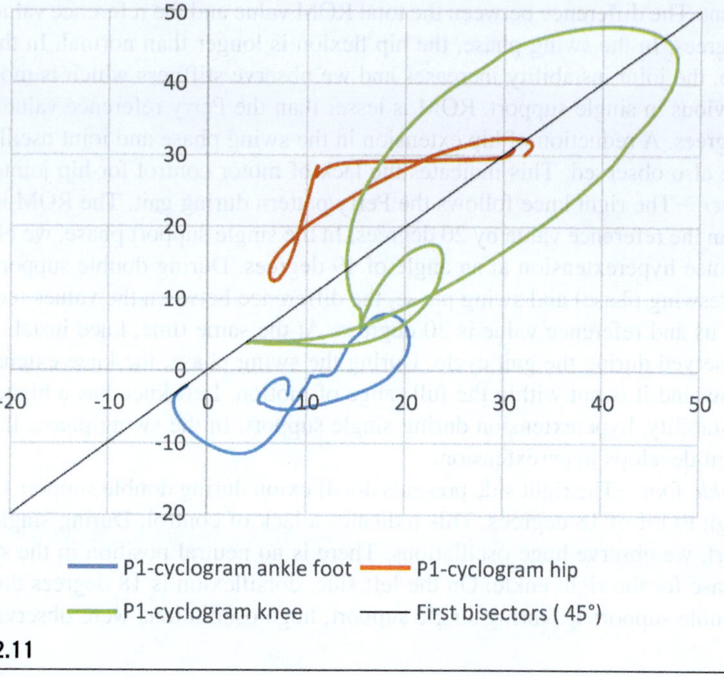

FIGURE 2.11

Cyclogram for symmetry.

5.2 Case study 2

Patient 2 (P2), male, 56 years old, weight 83 kg, height 1.85 m, body index mass (BMI) 24.25, MS, EDSS score 3. Clinical evaluation—gait disorders, balance disorders.

Kinematic parameters—The results regarding kinetic parameters (range of motion) are presented in Table 2.4, and the average values of indices (RI, SI, GA) are shown in Table 2.5.

In accordance with the reference values for ROM, we observe that the ROM is only 82.52% from normal ROM for hip. The ROM is 80.78% from normal ROM for knee, and the ROM is 73.92% from normal ROM for ankle.

A gait asymmetry, especially step high is observed. With respect to SI, we can observe an asymmetry for step high and step length. The kinetic and kinematic parameters have symmetry but there is no gait symmetry. In connection with the behavior of the lower limb joint during the gait cycle, we make an analysis of each joint during a gait cycle. The evaluation results are shown in Figs. 2.12–2.15. Table 2.6 shows the information about stance phase and swing phases.

Analysis of gait phases shows that the swing phase is close to the normal value. But, there is an increase in the stance phase. The evaluation of ROM for each joint is shown in Figs. 2.16–2.18. The results are presented in comparison with the reference values model proposed by Perry.

Interpretation of the results:

Hip—On the right side, we can observe hip instability and difficulty in movement. The difference between the total ROM value and the reference value is 15 degrees. In the swing phase, the hip flexion is longer than normal. In the left hip, the joint instability increases and we observe stiffness which is more obvious in single support. ROM is lesser than the Perry reference value by 10 degrees. A reduction of hip extension in the swing phase and joint oscillations are also observed. This indicates the lack of motor control for hip joints.

Knee—The right knee follows the Perry pattern during gait. The ROM is lesser than the reference value by 20 degrees. In the single support phase, we observe a knee hyperextension at an angle of 19 degrees. During double support 2 (preswing phase) and swing phase, the difference between the values recorded by us and reference value is 20 degrees. At the same time, knee instability is observed during the gait cycle. During the swing phase, the knee extension is slow and it is not within the full range of motion. Left knee has a higher instability, hyperextension during single support. In the swing phase, the patient develops hyperextension.

Ankle-foot—The right side presents dorsiflexion during double support 1, with a high ROM of 18 degrees. This indicates a lack of control. During single support, we observe huge oscillations. There is no neutral position in the stance phase for the right ankle. On the left side, dorsiflexion is 18 degrees during double support 1. During single support, huge oscillations were observed.

Table 2.4 Average values of range of motion (hip, knee, and ankle joints—sagittal plane).

	ROM (knee)	ROM (ankle)	ROM (hip)
P2	48.47 degrees	18.48 degrees	33.01 degrees

Table 2.5 Values of kinematic parameters and symmetry indices.

P2	Right	Left	RI	SI	GA
Step high (m)	0.186	0.241	−29.570	0.258	25.905
Step length (m)	1.357	1.42	−4.643	0.045	4.538
Speed gait (m/s)	1.18	1.23	−4.237	0.041	4.150
Cadence (steps/min)	52.17	51.95	0.422	0.004	−0.423

FIGURE 2.12

Heel strike (initial contact).

A comparison of the right versus left side clearly reveals that the left hip has a reduction of extension by 10 degrees more than the right hip. During the swing phase, both hips have a lot of oscillations with flexion/extension on the left side and extension/flexion on the right side. Regarding the knee, the main aspect is the hyperextension of the left knee in single support and a decrease of knee

FIGURE 2.13

Stance phase.

FIGURE 2.14

Preswing (toe off).

FIGURE 2.15

Swing phase.

Table 2.6 Components of gait cycle phases.

	Stance phase			Swing phase
	Doble support (% from gait cycle)	**Single support (% from gait cycle)**	**Doble support (% from gait cycle)**	**Single support (% from gait cycle)**
P2 right limb	14	36	12	38
P2 left limb	14	36	14	36

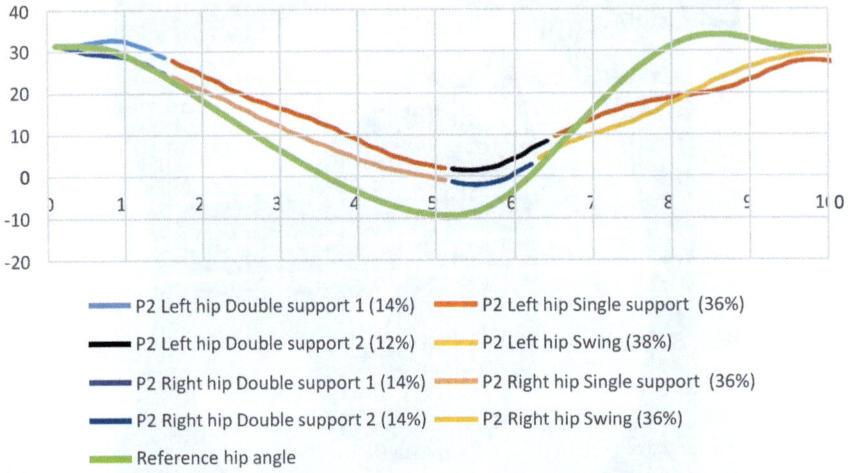

FIGURE 2.16

ROM for hip joints—right and left lower limbs.

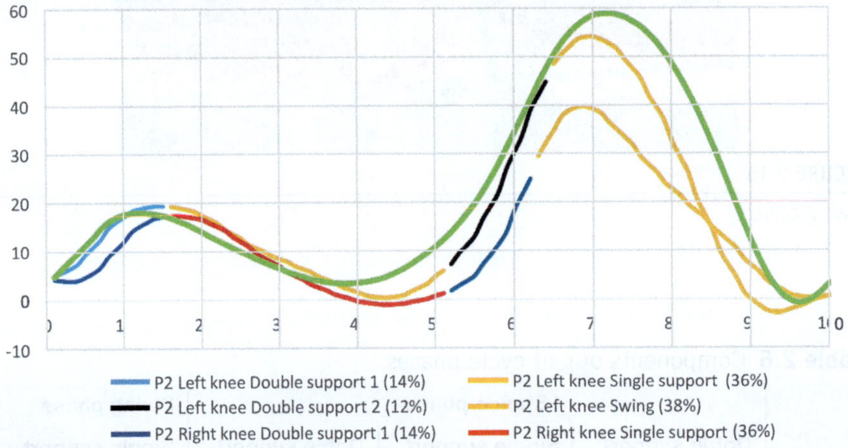

FIGURE 2.17

ROM for knee joints—right and left lower limbs.

flexion during the swing phase. Increase of dorsiflexion is more for the right ankle during the stance phase. Ankle oscillations during single support and huge oscillations on the left side were observed during the swing phase. The cyclograms for two couplers such as knee–hip and knee–ankle foot are presented in Figs. 2.19–2.21.

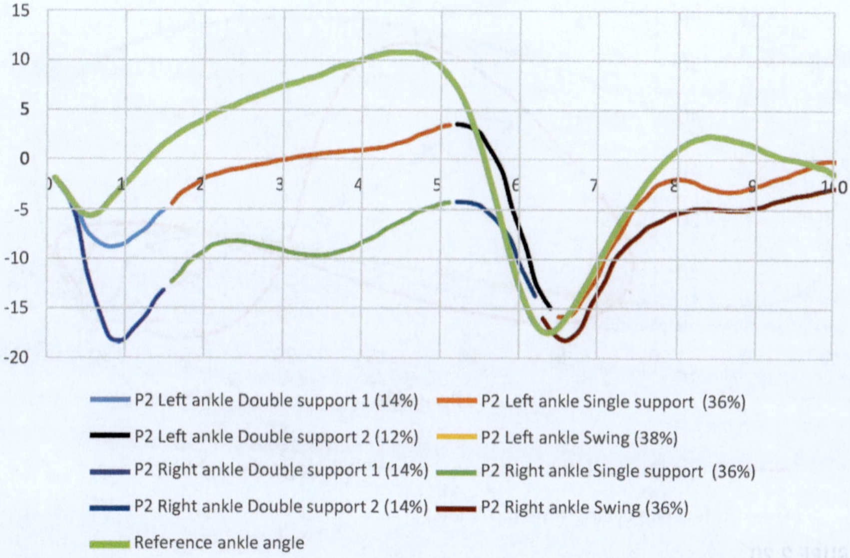

FIGURE 2.18

ROM for ankle joints—right and left lower limbs.

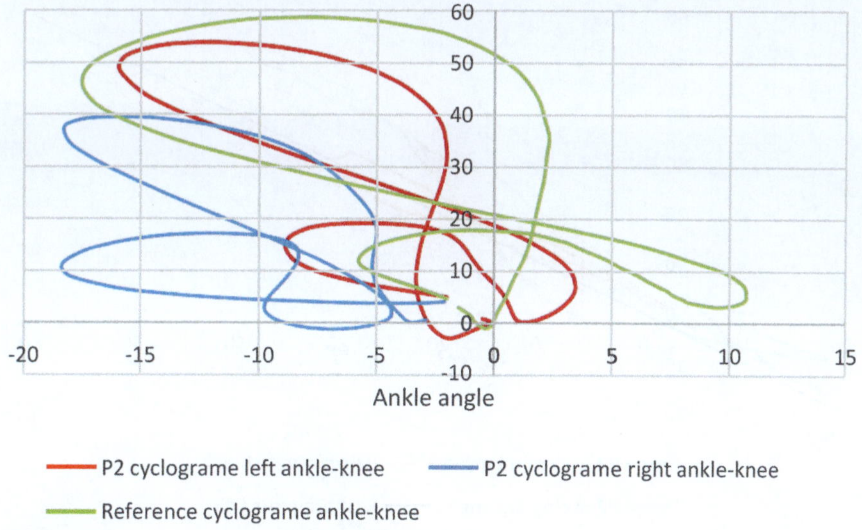

FIGURE 2.19

Cyclogram knee–ankle-foot/gait phases.

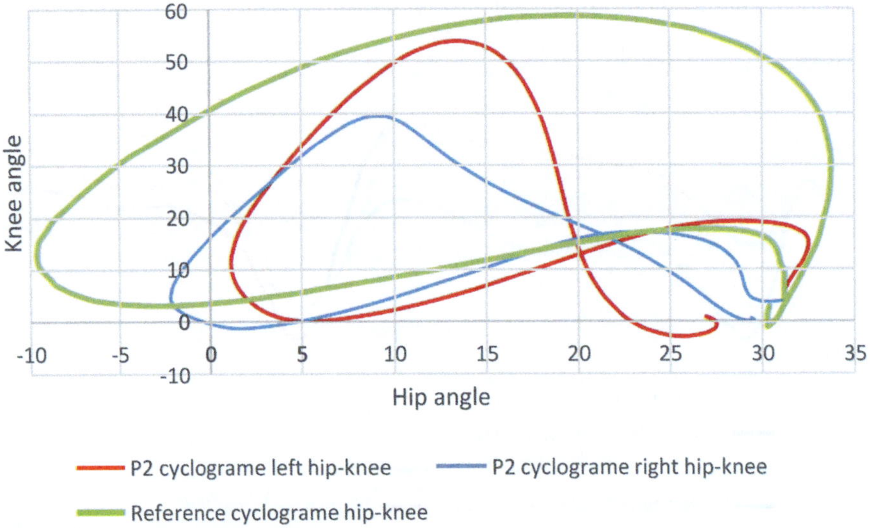

FIGURE 2.20

Cyclogram knee–hip/gait phase.

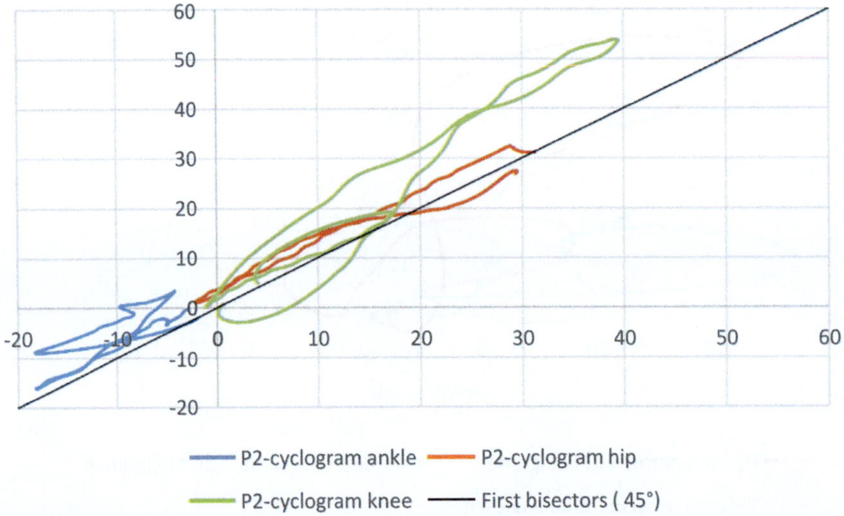

FIGURE 2.21

Cyclogram for symmetry.

Fig. 2.19 confirms the increase of dorsiflexion on the right side and the increase of knee flexion on the left side. Fig. 2.20 confirms the lack of right hip extension limit.

Using the cyclogram, we can analyze the symmetry. Fig. 2.21 shows the symmetry cyclogram, which shows the existence of symmetry for the hips and ankle foot.

5.3 Case study 3

Patient 3(P3), male, 13 years old, weight 52 kg, high 1.60 m, body index mass (BMI) 23.11, right hemiplegia (cerebral palsy). Clinical evaluation—gait disorders, balance disorders.

Kinematic parameters—The results of kinetic parameters (range of motion) are shown in Table 2.7, and the average values of indexes (RI, SI, GA) are shown in Table 2.8.

According to the reference values for ROM, it is observed that the ROM is only 58.42% from normal ROM for hip. The ROM is 85.65% from normal ROM for knee, and the ROM is 94.88% from normal ROM for ankle. From these results, we can conclude that the main problem is at hip joints level.

A gait asymmetry is observed, which is generated by step length. In terms of SI, we can observe a slow asymmetry, which means that the kinetic and kinematic parameters have symmetry. Regarding the behavior of the lower limb joint during the gait cycle, we make an analysis of each joint, and the results are presented in Figs. 2.22–2.25. Table 2.9 shows information about the stance phase and swing phases in the gait cycle.

Analysis of the gait phase demonstrates a decrease in the swing phase and an increase in double support especially for the left lower limb. This explains the role of the compensatory mechanism that tries to increase stability and balance.

The evaluation results of ROM for each joint are given in Figs. 2.26–2.28.

Interpretation of the results:

Hip—The right hip has high instability and the limit of extension is lesser than Perry value by 5%. Lack of extension is more visible in double support 2 (preswing). On the left hip, stiffness is observed along with oscillations and instability. The limit of extension is 20 degrees.

Knee—More instability is observed on the right side in double support 1 and single support. The knee has stiffness and is locked in flexion at 15 degrees. The limit of extension is very slow in the swing phase. Left knee, during single support, also has a limit of extension. This may be probably something like a compensatory mechanism. During double support 2 (preswing) and swing phase, the left knee behavior is similar to the Perry gait diagram.

Ankle—During double support 1, an increase in the dorsiflexion is observed on the right side. During single support, more oscillations with high frequency are observed, which decrease in double support 2 (preswing) and increase again during the swing phase. Dorsiflexion decrease on the swing phase and also plantarflexion decrease during the terminal swing phase.

Table 2.7 Average values of range of motion (hip, knee, and ankle joints—sagittal plane).

	ROM (knee)	ROM (ankle)	ROM (hip)
P3	51.39 degrees	23.72 degrees	23.37 degrees

Table 2.8 Values of kinematic parameters and symmetry indexes.

P3	Right	Left	RI	SI	GA
Step high (m)	0.178	0.168	5.618	0.058	−5.782
Step length (m)	0.902	1.042	−15.521	0.144	14.428
Speed gait (m/s)	0.56	0.62	−10.714	0.102	10.178
Cadence (steps/min)	36.92	35.85	2.898	0.029	−2.941

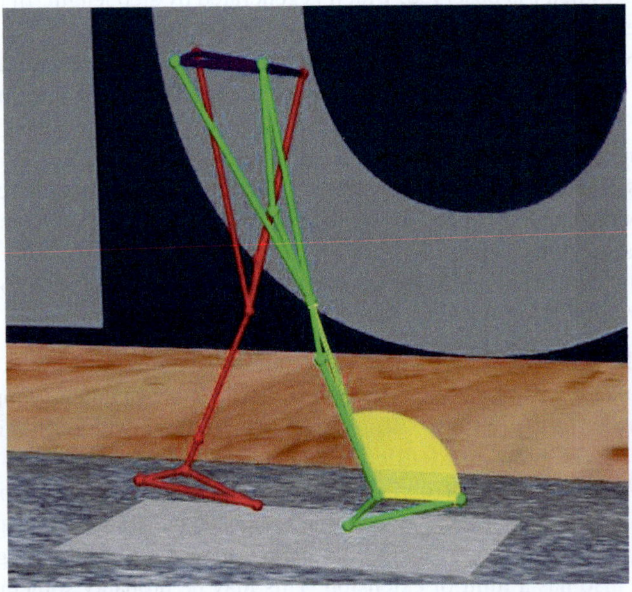

FIGURE 2.22

Heel strike (initial contact).

During double support 1, we observe a plantarflexion on the left ankle which seems to be like a compensatory mechanism because the right side has instability. This behavior is followed by dorsiflexion with oscillations during single support. Dorsiflexion increases during the swing phase and it is followed by plantarflexion. A comparison of left and right side results shows us that hip flexibility decreases

FIGURE 2.23

Stance phase.

FIGURE 2.24

Preswing (toe off).

FIGURE 2.25

Swing phase.

Table 2.9 Components of gait cycle phases.

	Stance phase			Swing phase
	Double support (% from gait cycle)	**Single support (% from gait cycle)**	**Double support (% from gait cycle)**	**Single support (% from gait cycle)**
P3 right limb	18	32	17	33
P3 left limb	21	31	17	31

with a different value of 30 degrees from the reference value. It is associated with instability and oscillations, which are predominant on the right side. Limit of knee extension is observed during the terminal swing. The diagrams show that the left side and the right side ankles are mirror images of each other. During the double support 2 (preswing), dorsiflexion increases on the left ankle. An analysis of gait is

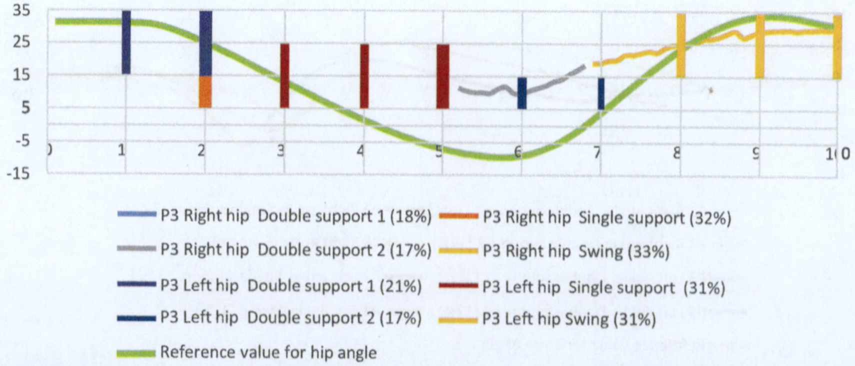

FIGURE 2.26

ROM for hip joints—right and left lower limbs.

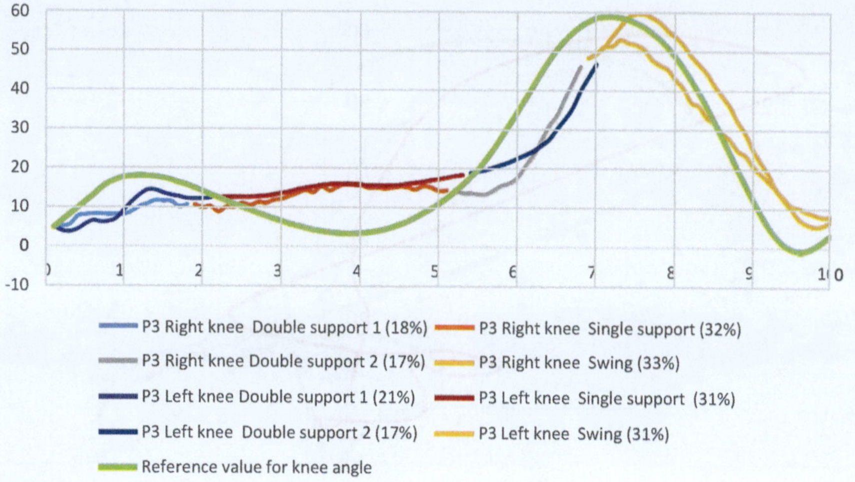

FIGURE 2.27

ROM for knee joints—right and left lower limbs.

useful to make the cyclograms for two couplers: knee–hip and knee–ankle foot. The cyclograms are presented in Figs. 2.29–2.31.

In the first cyclogram, we observe a high ROM for the left ankle in comparison with knee and reference values.

Fig. 2.30 demonstrates the limit of knee and hip extension during the gait cycle. The symmetry is presented in Fig. 2.31. It shows that the best symmetry is recording for knee joints and hip joints.

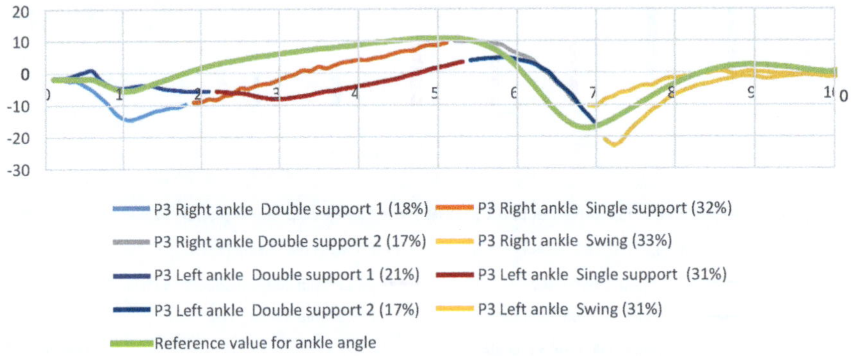

FIGURE 2.28

ROM for ankle joints—right and left lower limbs.

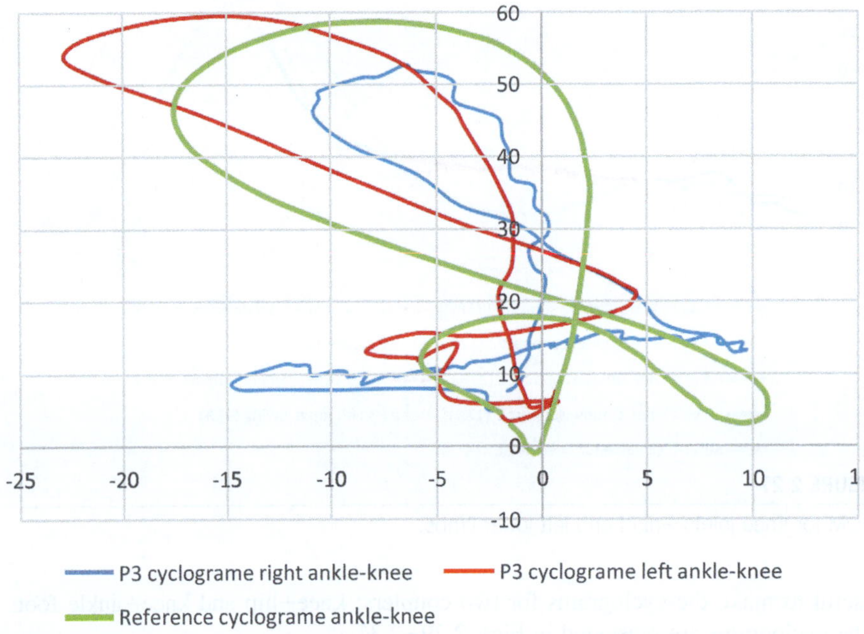

FIGURE 2.29

Cyclogram knee–ankle foot/gait phases.

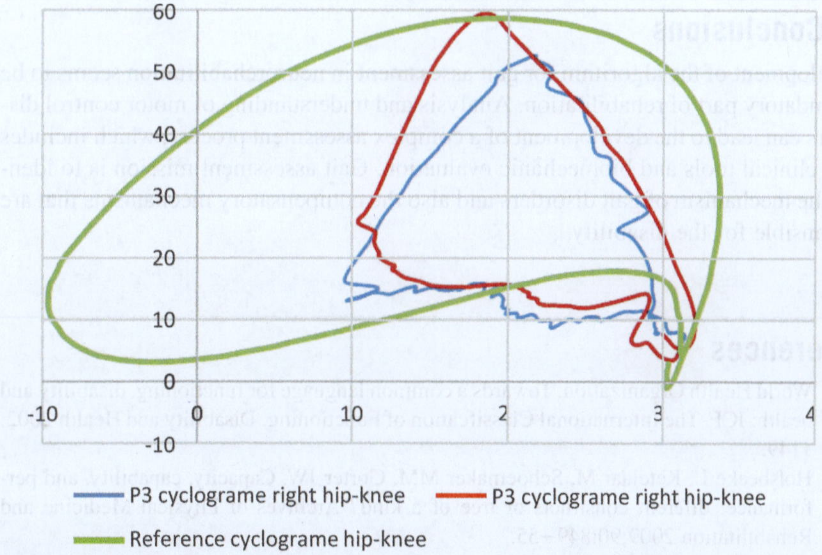

FIGURE 2.30

Cyclogram knee–hip/gait phase.

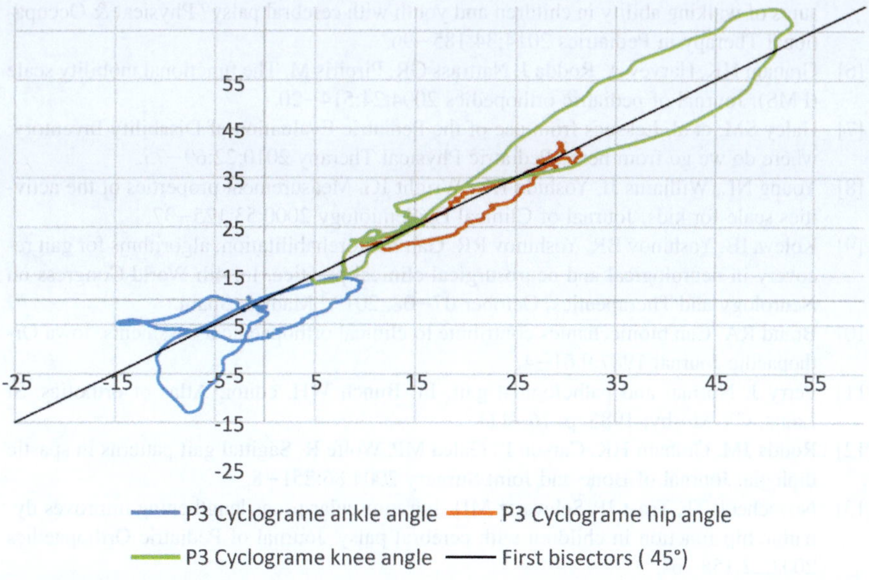

FIGURE 2.31

Cyclogram for symmetry.

6. Conclusions

Development of the algorithm for gait assessment in neurorehabilitation seems to be a mandatory part of rehabilitation. Analysis and understanding of motor control disorders can lead to the development of a complex assessment process, which includes both clinical tools and biomechanic evaluation. Gait assessment mission is to identify the mechanism of gait disorders and also the compensatory mechanisms that are responsible for the disability.

References

[1] World Health Organization. Towards a common language for functioning, disability and health: ICF. The International Classifcation of Functioning, Disability and Health 2002: 1149.

[2] Holsbeeke L, Ketelaar M, Schoemaker MM, Gorter JW. Capacity, capability, and performance: diferent constructs or tree of a kind? Archives of Physical Medicine and Rehabilitation 2009;90:849–55.

[3] Alotaibi M, Long T, Kennedy E, Bavishi S. The efficacy of GMFM-88 and GMFM-66 to detect changes in gross motor function in children with cerebral palsy (CP): a literature review. Disability & Rehabilitation 2014;36:617–27.

[4] Enright PL. The six-minute walk test. Respiratory Care 2003;48:783–5.

[5] Wilson NC, et al. How does the functional mobility scale relate to capacity-based measures of walking ability in children and youth with cerebral palsy? Physical & Occupational Therapy in Pediatrics 2014;34:185–96.

[6] Graham HK, Harvey A, Rodda J, Nattrass GR, Pirpiris M. The functional mobility scale (FMS). Journal of pediatric orthopedics 2004;24:514–20.

[7] Haley SM, et al. Lessons from use of the Pediatric Evaluation of Disability Inventory: where do we go from here? Pediatric Physical Therapy 2010;22:69–75.

[8] Young NL, Williams JI, Yoshida KK, Wright JG. Measurement properties of the activities scale for kids. Journal of Clinical Epidemiology 2000;53:125–37.

[9] Koleva IB, Yoshinov BR, Yoshinov RR. Gait neurorehabilitation: algorithms for gait recovery in neurological and neurosurgical clinical practice. In: 4th World Congress on Neurology and Therapeutics, October 07–08, 2019 | Madrid, Spain.

[10] Brand RA. Can biomechanics contribute to clinical orthopaedic assessments. Iowa Orthopaedic Journal 1987;9:61–4.

[11] Perry J. Normal and pathological gait. In: Bunch WH, editor. Atlas of orthotics. St Louis: CV Mosby; 1985. p. 76–111.

[12] Rodda JM, Graham HK, Carson L, Galea MP, Wolfe R. Sagittal gait patterns in spastic diplegia. Journal of Bone and Joint Surgery 2004;86:251–8.

[13] Novacheck TF, Trost JP, Schwartz MH. Intramuscular psoas lengthening improves dynamic hip function in children with cerebral palsy. Journal of Pediatric Orthopaedics 2002;22:158–64.

[14] Pirpiris M, Trivett A, Baker R, Rodda J, Nattrass GR, Graham HK. Femoral derotation osteotomy in spastic diplegia. Proximal or distal? Journal of Bone and Joint Surgery 2003;85:265–72.

[15] Pirpiris M. Single event multi-level surgery in spastic diplegia: comprehensive outcome analysis. Department of Paediatrics, University of Melbourne; 2002.

[16] Baker R. Pelvic angles: a mathematically rigorous definition which is consistent with a conventional clinical understanding of the terms. Gait & Posture 2001;13:1–6.

[17] Euler/Cardan sau Grood and Suntay Convention, Wu G, van der Helm FC, Veeger HE, Makhsous M, Van Roy P, et al. ISB recommendation on definitions of joint coordinate systems of various joints for the reporting of human joint motion–Part II: shoulder, elbow, wrist and hand. Journal of Biomechanics 2005;38:981–92.

[18] Baker R. ISB recommendation on definition of joint coordinate systems for the reporting of human joint motion- part I: ankle, hip and spine. Journal of Biomechanics 2003; 36:300–2.

[19] Schache AG, Baker R, Vaughan CL. Differences in lower limb transverse plane joint moments during gait when expressed in two alternative reference frames. Journal of Biomechanics 2007;40(1):9–19. https://doi.org/10.1016/j.jbiomech.2005.12.003.

[20] Leboeuf F, Baker R, Barré A, Reay J, Jones R, Sangeux M. The conventional gait model, an open-source implementation that reproduces the past but prepares for the future. Gait & Posture March 2019;69:235–41. https://doi.org/10.1016/j.gaitpost.2019.04.015.

[21] Davis RB, Ounpuu S, Tyburski D, Gage JR. A gait analysis data collection and reduction technique. Human Movement Science 1991;10:575–87.

[22] Kadaba MP, Ramakrishnan HK, Wootten ME. Measurement of lower extremity kinematics during level walking. Journal of Orthopaedic Research 1990;8:383–91.

[23] Kadaba MP, Ramakrishnan HK, Wootten ME, Gainey J, Gorton G, Cochran GVB. Repeatability of kinematic, kinetic, and electromyographic data in normal adult gait. Journal of Orthopaedic Research 1989;7:849–60.

[24] Ounpuu S, Gage JR, Davis RB. Three-dimensional lower extremity joint kinetics in normal pediatric gait. Journal of Pediatric Orthopaedics 1991;11:341–9.

[25] Ounpuu O, Davis RB, Deluca PA. Joint kinetics: methods, interpretation and treatment decision-making in children with cerebral palsy and myelomeningocele. Gait & Posture 1996;4:62–78.

[26] Pandy MG. Moment arm of a muscle force. Exercise and Sports Science Reviews 1999; 27:79–118.

[27] Gard SA, Childress DS. What determines the vertical displacement of the body during normal walking? Journal of Prosthetics and Orthotics 2001;13:64-7.

[28] Kerrigan DC, Riley PO, Lelas J, Della Croce U. Quantification of pelvic rotation as a determinant of gait. Archives of Physical Medicine and Rehabilitation 2001;82: 217–20.

[29] Gage JR. The treatment of gait problems in cerebral palsy. London: Mac Keith Press; 2004.

[30] Anderson FC, Pandy MG. Dynamic optimization of human walking. Journal of Biomechanical Engineering 2001;123:381–90.

[31] Heikkilä A, Sevander- Kreus N, Häkkinen A, Vuorenmaa M, Salo P, Konsta P, et al. Effect of total knee replacement surgery and postoperative 12 month home exercise program on gait parameters. Gait & Poature 2017;53:92–7.

[32] Ornettia P, Maillefert J-F, Laroche D, Morisset C, Dougados M, Gossec L. Gait analysis as a quantifiable outcome measure in hip or knee osteoarthritis: a systematic review. Joint Bone Spine 2010;77:421–5.

[33] Zagoa M, Sforza C, Bonardi DR, Guffanti EE, Galli M. Gait analysis in patients with chronic obstructive pulmonary disease: a systematic review. Gait & Posture 2018;61: 408–15.

[34] Fernando ME, Crowther RG, Lazzarini PA, Sangla KS, Buttner P, Golledge J. Gait parameters of people with diabetes-related neuropathic plantar foot ulcers. Clinical Biomechanics 2016;37:98–107.

[35] Kaczmarczyk K, Wit A, Krawczyk M, et al. Associations between gait patterns, brain lesion factors and functional recovery in stroke patients. Gait & Posture 2012;35:214–7.

[36] Kaczmarczyk K, Wit A, Krawczyk M, Zaborski J. Gait classification in poststroke patients using artificial neural networks. Gait & Posture 2009;30:207.

[37] Toro B, Nester CJ, Farren PC. The status of gait assessment among physiotherapists in the United Kingdom. Archives of Physical Medicine and Rehabilitation 2003;84: 1878–84.

[38] Harvey A, Gorter JW. Video gait analysis for ambulatory children with cerebral palsy: why, when, where and how. Gait & Posture 2011;33:501–3.

[39] Salbach NM, O'Brien KK, Brooks D, Irvin E, Martino R, Takhar P, et al. Reference values for standardized tests of walking speed and distance: a systematic review. Gait & Posture 2015;41(2):341–60.

[40] Mancini M, King L, Salarian A, et al. Mobility lab to assess balance and gait with synchronized body-worn sensors. Journal of Bioengineering and Biomedical Science 2011: 007.

[41] Favetta M, Romano A, Summa S, Colazza A, Minosse S, Vasco G, et al. Influence of sagittal pelvic attitude on gait pattern in normally developed people and interactions with neurological pathologies: a pilot study. Frontiers in Human Neuroscience August 4, 2022;16:797282. https://doi.org/10.3389/fnhum.2022.797282.

[42] Rudroff T, Proessl F. Effects of muscle function and limb loading asymmetries on gait and balance in people with multiple sclerosis. Frontiers in Physiology 2018;9:531.

[43] Pau M, Caggiari S, Mura A, Corona F, Leban B, Coghe G, et al. Clinical assessment of gait in individuals with multiple sclerosis using wearable inertial sensors: comparison with patient-based measure. Multiple Sclerosis and Related Disorders 2016;10:187–91.

[44] Pau M, Mandaresu S, Pilloni G, Porta M, Coghe G, Marrosu MG, et al. Smoothness of gait detects early alterations of walking in persons with multiple sclerosis without disability. Gait & Posture 2017;58:307–9.

[45] Morel E, Allali G, Laidet M, Assal F, Lalive PH, Armand S. Gait Profile Score in multiple sclerosis patients with low disability. Gait & Posture 2017;51:169–73.

[46] Goetschalckx M, Van Geel F, Meesen R, Triccas LT, Geraerts M, Moumdjian L, et al. Interlimb coordination performance in seated position in persons with multiple sclerosis: reduced amplitude over 6 min and higher coordination variability in persons with walking fatigability. Frontiers in Human Neuroscience October 20, 2021;15: 765254. https://doi.org/10.3389/fnhum.2021.765254.

[47] Farrell III JW, Merkas J, Pilutti LA. The effect of exercise training on gait, balance, and physical fitness asymmetries in persons with chronic neurological conditions: a systematic review of randomized controlled trials. Frontiers in Physiology 2020;11:585765.

[48] Mentiplay BF, Perraton LG, Bower KJ, Pua YH, McGaw R, Heywood S, et al. Gait assessment using the Microsoft Xbox one Kinect: concurrent validity and inter-day reliability of spatiotemporal and kinematic variables. Journal of Biomechanics 2015; 48(10):2166–70.

[49] Olney SJ, Richards C. Hemiparetic gait following stroke. Part I: characteristics. Gait & Posture 1996;4(2):136–48.

[50] Flansbjer UB, Holmbäck AM, Downham D, Patten C, Lexell J. Reliability of gait performance tests in men and women with hemiparesis after stroke. Journal of Rehabilitation Medicine 2005;37:75–82.

[51] Mazzoli D, Basini G, Prati P, Galletti M, Mascioli F, Rambelli C, et al. Indices of loading and propulsive ability in the gait of patients with chronic stroke with equinus foot deviation: a correlation study. Frontiers in Human Neuroscience January 14, 2022;15:771392. https://doi.org/10.3389/fnhum.2021.771392.

[52] Tabard-Fougère A, Rutz D, Pouliot-Laforte A, De Coulon G, Newman CJ, Armand S, et al. Are clinical impairments related to kinematic gait variability in children and young adults with cerebral palsy? Frontiers in Human Neuroscience March 2, 2022; 16:816088. https://doi.org/10.3389/fnhum.2022.816088.

[53] Chiou II, Burnett CN. Values of activities of daily living: a survey of stroke patients and their home therapists. Physical Therapy 1985;65:901–6.

[54] Von Schroeder HP, Coutts RD, Lyden PD, et al. Gait parameters following stroke: a practical assessment. Journal of Rehabilitation Research and Development 1995;32: 25–31.

[55] Coca-Tapia M, Cuesta-Gomez A, Molina-Rueda F, Carratalà-Tejada M. Gait pattern in people with multiple sclerosis: a systematic review. Diagnostics 2021;11:584.

[56] Spain RI, St George RJ, Salarian A, Mancini M, Wagner JM, Horak FB, et al. Body-worn motion sensors detect balance and gait deficits in people with multiple sclerosis who have normal walking speed. Gait & Posture 2012;35:573–8.

[57] Longworth LA, Chlosta S, Foucher KC. Inter-joint coordination of kinematics and kinetics before and after total hip arthroplasty compared to asymptomatic subjects. Journal of Biomechanics 2018;72:180–6.

[58] Park JH, Lee HH, Cho J, Kim I, Lee J, Jang SH. Effects of knee osteoarthritis severity on inter-joint coordination and gait variability as measured by hip-knee cyclograms. Scientific Reports 2021;11:1789.

[59] Field-Fote EC, Tepavac D. Improved intralimb coordination in people with incomplete spinal cord injury following training with body weight support and electrical stimulation. Physical Therapy 2002;82:707–15.

[60] Awai L, Curt A. Intralimb coordination as a sensitive indicator of motor-control impairment after spinal cord injury. Frontiers in Human Neuroscience 2014;8:148.

[61] Conner BR. Relationship between ankle function and walking ability for children and young adults with cerebral palsy: a systematic review of deficits and targeted interventions. 2022. p. 165–8. https://doi.org/10.1016/j.gaitpost.2021.10.024.

[62] Zhou JB. Neurologic correlates of gait abnormalities in cerebral pals: implication for treatment. Frontiers in Human Neuroscience 2017;11:103. https://doi.org/10.3389/fnhum.2017.00103.

[63] Grigoriu AL. Characteristics of newly acquired gait in toddlers with unilateral cerebral palsy: implications for early rehabilitation. Annals of Physical and Rehabilitation Medicine 2021;64(3):101333. https://doi.org/10.1016/j.rehab.2019.10.005.

[64] Elnaggarab R. Relationship between transverse-plane kinematic deviations of lower limbs and gait performance in children with unilateral cerebral palsy: a descriptive analysis. Gait & Posture 2021:224–8. https://doi.org/10.1016/j.gaitpost.2020.05.003.

[65] Tofthagen C, Overcash J, Kip K. Falls in persons with chemotherapy- induced peripheral neuropathy. Supportive Care in Cancer March 2012;20(3):583−9. https://doi.org/10.1007/s00520-011-1127-7. Epub March 5, 2011.

[66] Davis RB, Õunpuu S, Tyburski D. A gait analysis data collection and reduction technique. Human Movement Science 1991;10:575−87.

[67] Seliktar R, Mizrahi J. Some gait characteristics of below-knee amputees and their reflection on the ground reaction forces. Engineering in Medicine 1986;15:27−34. https://doi.org/10.1243/emed_jour_1986_015_009_02.

[68] Andre R, Stimmel S. Prosthetic alignment effects on gait symmetry, a case study. Clinical biomechanics 1990;5(2):88−96.

[69] Sadghi H, Allard P, Prince F, Labeller H. Symmetry and limb dominance in able-bodied gait: a review. Gait & Posture 2000;12(1):34−45.

[70] Plotnik M, Giladi N, Hausdorff J. A new measure for quantifying the bilateral coordination of human gait, effects of aging and Parkinson's disease. Experimental Brain Research 2007;181(4):561−70.

[71] Kutilek P, Viteckova1 S, Svoboda Z, Smrcka P. Kinematic quantification of gait asymmetry in patients with peroneal nerve palsy based on bilateral cyclograms. Journal of Musculoskeletal and Neuronal Interactions 2013;13(2):244−50.

[72] Blazkiewicz M, Wiszomirska I, Andrezej W. Comparison of four methods of calculating the symmetry of spatial-temporal parameters of gait. Acta of Bioengineering and Biomechanics 2014;16(1). https://doi.org/10.5277/abb140104.

CHAPTER 3

Deep learning method for adult patients with neurological disorders under remote monitoring

K. Kathiresan[1], T. Preethi[1], N. Yuvaraj[1], S. Karthic[1], K.R. Sri Preethaa[1,2]

[1]*Department of Computer Science and Engineering, KPR Institute of Engineering and Technology, Coimbatore, Tamil Nadu, India;* [2]*Department of Robot and Smart System Engineering, Kyungpook National University, Daegu, Republic of Korea*

1. Introduction

As we age, our brains also deteriorate over time. The majority of cases of Alzheimer's disease occur in individuals over the age of 65, making aging the primary risk factor for the disease. However, it is important to note that cases of Alzheimer's have been reported in individuals much younger than 65, indicating that the disease is not solely a result of aging and it is not a normal part of the aging process [1]. Dementia is a broad term that encompasses a range of cognitive impairments, including declines in memory and reasoning abilities. The majority of dementia cases, 60%–80%, are caused by Alzheimer's disease, which progressively diminishes a person's ability to think and act independently [1].

The number of persons with dementia is currently 50 million, and it is predicted that this number will double every 20 years to reach 152 million by 2050 [2]. Scientists are working against the clock to develop viable medicines to avert and treat the disease's symptoms as a result of this worrying rise in sufferers [3]. Being a progressive condition, it begins with mild memory loss and worsens with time to the point when the patient is unable to talk or reply to external stimuli [4]. Patients with Alzheimer's disease may either neglect to eat, leading to undernourishment or eat too frequently throughout the day, leading to overeating [5].

Dehydration and improper eating are considerably more common in those with Alzheimer's disease, and they even neglect to eat or drink even when they are hungry or thirsty. As a result of the disorientation, they experience from being unable to rely on short-term memory, patients are more likely to eat more or more frequently [6]. Thus, it becomes nearly difficult to live freely, necessitating the requirement for a carer. Therefore, an approach that observes the behaviors of patients to ascertain when they eat and drinks as well as notifies both the patient and their carer of the critical timing when the sufferer should have water or food to prevent certain consequences is needed.

Computational Intelligence and Deep Learning Methods for Neuro-rehabilitation Applications
https://doi.org/10.1016/B978-0-443-13772-3.00008-X

- To allow carers to keep an eye on the patient's activities, the existing systems include person monitoring systems, reminders for chores like taking medication, and video surveillance systems. These systems are made to aid those who have dementia and Alzheimer's. As examples, consider the following:
- Reminder applications: (a) The iOS app "Pill Reminder" serves as a reminder for the patient to take his prescription at the appointed time. The patient can manually log in to the app to confirm when the medication was taken. These notifications are established by a carer. It's finished! A reminder app for iOS (b) enables the carer to arrange a list of tasks for the patient to accomplish during the day and check them off as they are finished [7].
- Monitoring systems: Real-time video streaming to caregivers is made possible by modern monitoring methods, but the caregiver must constantly monitor the stream to observe the patient's behavior. "Foscam" security cameras enable this and also allow for audio communication between the caregiver and patient.
- "Y-cam" cameras have the feature of motion-activated monitoring, and other options for real-time security cameras include Baby Cam and Netgear Arlo [8].

Nowadays AI-based products are now present in the market that are designed to actively support AD patients in real-time with crucial parameters like food or dehydration. The majority of the implementations talk about task lists, perimeter monitoring, and memory stimulation. The following cures were found as a result: (1) Constant Therapy, a tablet, and a smartphone app that offers cognitive, linguistic, and acoustic therapy. It was developed to support people suffering from Alzheimer's and dementia and also those who were getting cures for head damage like strokes. (2) Alz Calls is a chatbot created for individuals with Alzheimer's disease who need social interaction, have trouble adjusting to new surroundings, or regularly enquire about their family.

The members can post their voices, and provide an image that will appear for the patient to realize and react to typical scenarios when the caretaker is unable to interact with the patient. (3) The amusement app is developed to provide interaction which includes about 1000 simple quiz questions incorporated with a substantial ocular component. (4) A program called timeless uses facial recognition technology based on artificial intelligence to assist Alzheimer's patients in recognizing people, recalling events, and staying in touch with friends and family. (5) MindMate is a thorough tool for dementia sufferers. To reduce a few of the consequences of cognitive decline, this program was designed based on the most recent examination of dementia. It uses mentally demanding games, nutritional advice, and exercise. (6) With the assistance, A caretaker can send reminders to a friend or relative who has dementia using the app MemoClock. It is possible to establish daily repeating reminders. (7) The application's 40 daily tasks are listed in the app It's Done! to aid dementia patients in remembering what has to be done. The program allows users to make their task lists to keep track of specific jobs and check them off as they are finished. (8) Ode is an application that simultaneously creates three all-natural food scents to whet patients' appetites as they eat. I Human Link is a kit that includes an app that is downloaded to family members' phones, a watch with a built-in SIM, and a watch that the Alzheimer's sufferer wears. Moreover, the watch enables users to dial saved preferred numbers using a phone function. Another option is to designate safety zones or a boundary around which the individual performs his or her activity. If the patient exits this area, the family will be notified right away.

Among the technologies and software developed to assist the public with Alzheimer's, none of the methods are available to track the actions of the patient to restrain dehydration, starvation, or overeating. Caregivers cannot stop closely monitoring the activities of the patients they are responsible for because no technology can recognize activities generated for them automatically using artificial intelligence. Both persons with Alzheimer's disease and those Without it, human behavior can still be observed and detected thanks to the present surveillance technology. individuals who care for those with Alzheimer's disease; The patient and carer are informed when the sufferer needs to consume or stop consuming food or water. The recording is delivered in real time, and the recording of the monitored person's eating or drinking is preserved in a database. The Alzheimer's patient can reclaim his confidence by not being required to reside in the locations of his carers. Because the artificial vision-based surveillance system eliminates the need to constantly monitor the patient's activity, the carer also benefits from technology in terms of the patient's requirements and health. Furthermore, if the carer so chooses to observe the patient's activities, he can finish it virtually with just a smart device consisting of a web browser and internet access.

In this article, artificial vision-related models and algorithms are investigated, sought for, trained, and applied to categorize photos, locate objects in images, and detect human movement in movies.

Adding to the Healthcare 4.0 paradigm, which is discussed in Ref. [9], comes the new study. This paradigm states that data-driven, patient-centered health systems will be realized, enabling the use of cutting-edge sensors to provide individualized assistance. The authors contend that sophisticated algorithms and enabling technology can aid intellectually disabled persons with diseases like Alzheimer's. They concentrate on both emotion recognition and sensory movement data in their research. A different study [10] tries to give people with assistive intelligence devices who have perceptual issues a platform. The authors stated that wearable IoT device development has been impeded by the lack of computing power in these devices up to this time. However, the suggested system's portable computational capabilities have advanced to the point where real-time operations are now practicable. This study significantly contributes to the Healthcare 4.0 ecosystem in several ways, including the following: The suggested methods can be used in specialized systems that are not necessary for treating Alzheimer's disease, and it is demonstrated that present portable hardware is capable of processing artificial intelligence conclusions in almost real time. It focuses on a specific and important activity (Alzheimer's patients' eating habits) that are not given significant importance in other research works.

2. Related works

As a result of ongoing efforts by scientists in the area to develop tools that are capable of carrying out any activity a person can, artificial intelligence has developed to the point where it is currently. With personal assistants on smart devices like Siri, Alexa, and Google Assistant, this business already helps people in their daily lives and, in some cases (such as the usage of dangerous chemicals and labor in huge production facilities), even goes beyond what is physically possible [11]. The use of artificial intelligence in daily chores has advanced significantly during the last 10 years. Artificial intelligence has grown as a result of constant research by scientists in the field to create technologies that are capable of performing any task a person can. Artificial intelligence has advanced to the level it is at now. This industry already makes life easier for individuals with smart devices, and in some instances (like the use of hazardous chemicals and labor in massive production facilities) even goes beyond what is physically possible [11]. Artificial intelligence applications have advanced dramatically during the previous 10 years for routine tasks. The creation of robots with similar human-like perception, comprehension, interpretation, determination, and response to external stimuli is the aim of artificial intelligence research and development. The aim of artificial intelligence development is for machines to sense, comprehend, interpret, determine, rationalize, discern, and react to environmental stimuli in the same ways that humans can [11].

Several studies have employed deep learning (DL) techniques, including recurrent neural networks, deep neural networks, and convolutional neural networks (CNN), to investigate Alzheimer's disease [12–14]. Some authors [15] developed a nonlinear metric learning method using a stacked denoising sparse auto-encoder in a deep-network-based feature fusion method to combine longitudinal and cross-sectional features derived from magnetic resonance brain images to enhance biomarker identification for Alzheimer's disease. Other researchers [16] proposed a deep belief network (DBN), a prediction model-based algorithm, as a method for classifying Alzheimer's disease based on structural modalities. According to some studies [17], a pretrained AlexNet model that extracts deep features from data and an algorithm for determining the stage of Alzheimer's illness and a pretrained AlexNet model is used to extract deep features from data, and an algorithm is used to determine the stage of Alzheimer's illness. The pretrained AlexNet model also removes the initial layers and deep features from the convolutional neural network pretrained model (CNN). Certain publications [18] used a weakly supervised learning-based deep learning (DL) architecture (ADGNET) for concurrent image categorization and picture reconstruction when there are few annotated pictures. This architecture is composed of a primary network with a focus mechanism and a specific task network. Recent research [19] found that by combining global shape representations with hippocampus segmentation, a deep convolutional network model, DenseCNN2, can classify Alzheimer's disease.

Following a thorough methodology for Alzheimer's disease diagnosis, a few pieces of literature [20] discovered that CNN outperforms other deep models in classification tasks, whereas [21,22] discovered that DL techniques are superior to the traditional machine learning methods and shallow neural network architecture. Recurrent neural networks are utilized in a recent computerized evaluation method for Alzheimer's disease that is dependent on audio [23]. Various researchers have also improved automatic Alzheimer's disease prediction by utilizing a transfer learning technique without the usage of expert-defined linguistic characteristics and a small, focused speech dataset [24]. It might be argued that the majority of supervised learning techniques were used to try and quickly identify, detect, and anticipate the early symptoms of Alzheimer's disease. The main focus is to improve customized assistance for Alzheimer's disease sufferers by utilizing cutting-edge technologies based on computer vision. A recent study [9] proposes an Internet of Things-based support system for individuals with Alzheimer's disease that may warn people about bad decisions and offer to reinforce and support help to promote the appropriate decisions. The architecture is built on deep learning and contains both an IoT-based help mechanism and an ensemble approach for tracking abnormalities. An Internet of Health system built on deep learning that benefits both the patient and carer was created because eating and drinking can be challenging for those with Alzheimer's disease. A viable way for delivering real-time, individualized support may be a framework that could detect food and drinks and, as a result, track human activities connected to eating and drinking to verify that they are done appropriately.

As per the literature [25], the importance and impact of food on people's lives have led to the development of innovative computer vision methods for automatically detecting food, identifying food for automatic harvesting, and evaluating food quality. These methods are particularly useful for patients with conditions related to food, such as obesity, diabetes, or Alzheimer's disease. Authors [26,27] have used categorization on food photographs to train the model and evaluate meal quality. They then extracted a variety of attributes from the generated segments and implemented SOM (Self-Organizing Feature Map) to enhance classification performance. A monitoring system was developed by several authors [28]. Using two CNNs to identify meals and recognize foods, calorie counting using a deep learning model involving a library of food photos was successfully tested for identifying the food intake using real-time images captured from a mobile camera a recent study [29], a multitask fully convolutional network for picture segmentation and RGB-D image pairs taken before and after the meal were used to construct an AI-based monitoring system for nutritional intake by hospitalized patients. It follows that food can be exactly and precisely measured successfully detected and classified using deep learning algorithms. This can then be used as the basis for developing fresh patient health aid solutions that promote wise choices, particularly for those with Alzheimer's disease. Other researchers [30] contend that a disadvantage of utilizing deep learning in the study of Alzheimer's disease is that it can be difficult to correct bias When network complexity is too high to ensure transparency and repeatability, it might occur in the system. Nevertheless, by gathering a lot of imaging data and examining the connections between deep learning and characteristics, this problem might be rectified. The application of deep learning to identify Alzheimer's disease may be constrained, according to other studies [31], because a thorough explanation of the depth model and the benchmarking platform is necessary to find the best combination of several indications.

Additionally, the majority of studies on deep learning for recognizing the consumption of food and beverages rely on experimental work carried out in extremely constrained environments, which should be seen as the main limitation. The lack of publicly accessible datasets may also provide a substantial barrier to the development of patient aid systems and deep learning applications for disease detection, according to recent research [32].

2.1 Modern artificial intelligence

- The advancements in artificial vision, language recognition, and self-driving cars today can be traced back to the breakthroughs of 2012, which included a neural network developed by researcher Geoffrey Hinton and his colleagues and used by them to win an international artificial vision competition [33]. These elements have altered how people now see artificial intelligence [11,21].
- Artificial intelligence encompasses various subfields, such as natural language processing (the ability to analyze and understand natural language information in documents), expert systems (which utilize a knowledge base of rules to solve complex issues), computer vision (the capability of machines to perceive and interpret their environment), robotics (the development, programming, and creation of robots), and machine learning (the ability of a machine to learn, with deep learning as a specific subfield of machine learning).
- Speech recognition and process automation (which employ robot automation and artificial intelligence) (deals with speech and voice recognition)
- When creating an artificial intelligence project, it is frequently necessary to use a variety of subdomains due to their interconnectedness.

2.2 Deep learning

Artificial neural networks were once unable to be effectively trained in real-world settings due to limitations in computer power and dataset size, even as more neurons were added to the network. However, this has changed with the advent of deep learning and the research of Geoffrey Hinton. As a result, areas such as artificial vision and voice recognition have seen significant advancements, as the layers of neurons can now be properly trained [33].

2.3 Convolutional neural networks

CNN are tailored for analyzing visual data in 2D, 3D, and similar formats. They are distinct from traditional neural networks in that they incorporate convolutional layers and employ pooling layers for image compression. These networks produce the output through fully connected layers (where all neurons are connected to each other). Among the intriguing uses of artificial vision include picture classification, which involves classifying an entire image after taking it all in.

Finding, recognizing, and framing one or more things in an image is called object detection. Finding and recognizing the pixels that make up an image's objects is a part of image segmentation.

"Video activity recognition" is the process of analyzing a whole video and grouping the outcomes into a single category. Locating and identifying the location of the video activity, deep learning can be used to precisely diagnose conditions like skin imperfections, neurological abnormalities, and chronic illnesses, according to a recent study [34]. So, employing a convolutional neural network, a recent study [21,32] based on 30 movies collected from a private eye facility successfully identified dry eye sickness from tear film breakdown time. Four classes—normal, breakup, blink, or noise—are established when converting six videos into 100 frames for testing. 600 frames from 14 training films, or 70% of the available footage, are used to train the convolutional neural network. The researchers [32] accuracy, sensitivity, specificity, and correlation coefficient were all able to reach 83%. Other authors [35] also provided evidence on how the blend of deep learning with optical coherence tomography (OCT) technology showed improved diagnostics and made it possible for the earlier identification of retinal infections. Some researchers [36] think that cutting-edge deep learning algorithms could be an alternative to present image processing techniques [37,38] given that this tendency has been observed in other medical specializations including gastrointestinal disorders and radiology. A different study [12,39] proposes a technique for calculating a meiboscore by first extracting features from photos of meibography using a specific sort of unsupervised CNN and then using a K-nearest neighbors model. While training the ant-lion optimized single feedforward networks, the deep fused features-based reliable optimized networks that some authors [40] proposed method combines saliency maps and convolutional layers to improve segmentation and feature extraction while reducing the overfitting problem. The whale optimization algorithm is used to adjust the hyperparameters of pretrained networks to enhance prediction accuracy. The results surpass those of AlexNet, Resnets-50, VGG-16, DenseNet, Resnets-100, and Genesis models, with accuracy, precision, recall, and F1-score values of 98.5%, 99.0%, 98.8%, and 99.1% respectively. A controlled CNN classifier developed by other researchers [41] to identify likely adenocarcinoma and squamous cell carcinomas in lung cancer patients had a subpar prediction performance of 71%.

2.4 Data source

With modern computing power, growing datasets, and a massive number of examples needed to create the experience that the program learns from, artificial neural networks can now be deployed. The amount of data for image or video datasets has increased because the majority of people own a camera or other video recording equipment (e.g., a smartphone, camera, and webcam). The ability to quickly access social media platforms and the internet, where a variety of images and videos may be shared, makes it possible to gather the necessary digital data. The picture classification algorithms utilize the IMDB-WIKI dataset [42–44], Furniture Detector [45], and Food-101 [46] datasets. The aforementioned databases either include images of food or are often seen considering that classifying photos as either food or nonfood is useful (furniture and people).

2.5 Image classification

CNNs were constructed from scratch using the transfer learning method to classify the photographs. The CNN was constructed using the VGG-16 architecture for image classification, which was trained on the ImageNet dataset. This architecture was selected after evaluating different models, including those developed from scratch and transfer learning models such as MobileNet, InceptionV3, and VGG16, for performing binary classification for food identification (food class). Different datasets were also examined for various categories of image classification from the images.

3. Proposed methodology

3.1 Image classification

Three types of AI models were used in the creation of a deep-learning-based system for assisting Alzheimer's patients: image categorization, item identification, and human behavior recognition. The capacity to distinguish between things that are for consumption and those that are not is a crucial system capability that required image classification analysis. The study put forth the possibility that models with exceptional classification abilities (over 80%) may be created utilizing the present deep learning approaches. The built model's later shown accuracy of 96% was high enough to be acknowledged as helpful. The picture categorization modeling procedure was carried out using Python programming and the Jupyter Notebook environment. The software library used for doing numerical computations was called NumPy. Utilizing Glob, Matplotlib, and Scikit-learn (often abbreviated as sklearn), for getting a list of file paths for each image in the dataset from the supplied directory, presenting graphs, computing the model's accuracy on the test subset, and creating the confusion matrix. An artificial neural network simulation is carried out using Keras. The dataset is split into training and testing subsets, with each having 250 images and the training subset having 1000 images for each class—see Table 3.1.

In light of this tactic, the model was quickly trained using the transfer learning technique with a small dataset to enhance its performance on the test set. As a result, the VGG16 convolutional neural network model served as the foundation for training the new model—look at Fig. 3.1. The 14 million image ImageNet dataset

Table 3.1 Binary classification dataset.

Dataset details				Train		Test	
Total images	Image size	Food	Not food	Food	Not food	Food	Not food
2500	227 × 227	1250	1250	1000	250	1000	250

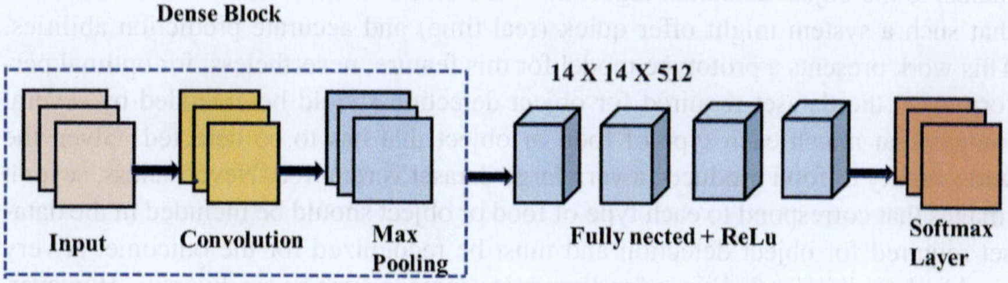

FIGURE 3.1

Convolutional blocks in VGG16 architecture.

was used to train this model over many weeks. With a 92.7% accuracy rate, it was among the top five models evaluated using the ImageNet dataset [47].

Five Fig. 3.1 depicts the VGG16 architecture, which is made up of convolutional blocks. The first two blocks of the model consist of two convolutional layers and a max pooling layer. The next three blocks include a max pooling layer and three convolutional layers. The accuracy of the model was evaluated using a confusion matrix (Fig. 3.2) and the results were obtained from testing 500 images: 241 true positives (model correctly identified the positive class as "food"), 243 true negatives, seven false positives (model correctly recognized the negative class as not being food), and nine false negatives (model incorrectly predicted the positive class) [48].

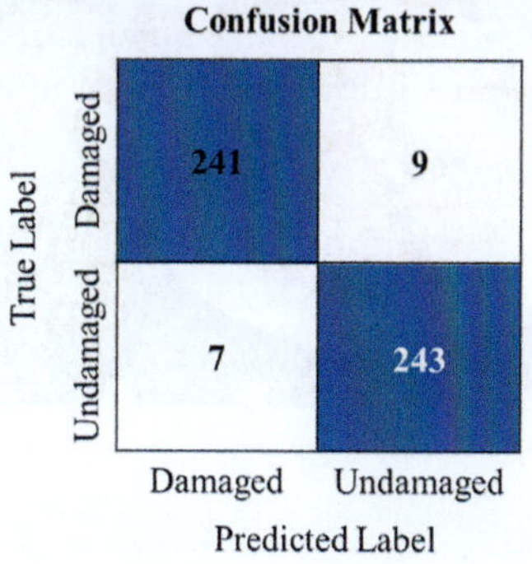

FIGURE 3.2

Confusion matrix.

A properly trained model that works well is demonstrated by the confusion matrix, which reveals that fewer incorrect predictions were made than were made overall.

3.2 Object detection

In addition to distinguishing between food and nonfood things, The system can accurately identify the names of several different objects, such as forks or knives, thanks to the object detection algorithms. If trained on sufficient data, it is believed that such a system might offer quick (real time) and accurate prediction abilities. This work presents a prototype model for this feature; nevertheless, for optimal performance, the dataset required for object detection should be extended by several images that match each type of food or object that has to be detected. Given the wide variety of food products, a very large dataset is required. Nevertheless, several images that correspond to each type of food or object should be included in the dataset required for object detection and must be recognized for the outcome. A very sizable dataset is needed since food products come in such a vast diversity. However, with the correct technology, the models' training may be finished rapidly (high-end video processors).

Detecting objects in a photograph requires being able to identify and locate them. The objects of interest on the image used for the object detection process will be drawn in a frame (Fig. 3.3). The convolutional neural network was trained using

FIGURE 3.3

Annotating images.

the SSD ResNet50 V1 FPN 640 × 640 (RetinaNet50) architecture and the transfer learning method to identify objects in the images. In addition, a speedier R–CNN ResNet50 V1 640 × 640 model that had been pretrained on the Coco2017 dataset was evaluated. The "Detection Model Zoo" in the TensorFlow two collection of models, which is accessible for use without charge [27], contains the two pretrained models. Annotating a collection of images with one's information, which includes Additional, photographs of interesting objects created. The photographs were personally found, selected, and using the labeling tool to add annotations (Fig. 3.3) [49].

Parts of object detection include both object recognition and object identification. A frame containing the recognized objects of interest will be constructed on the output image of the object detection method.

The ResNet50 V1 FPN 640 × 640 (RetinaNet50) SSD model displays the network architecture of the ResNet50, which has 50 convolutional layers and is used for object detection in images (see Fig. 3.4). The supported images have 50 convolutional layers, one maximum pooling layer, and one average pooling layer. They are 640 × 640 pixels in size (for image compression). The last layer, according to the classification of the 1000 class categories from the ImageNet dataset, as represented by the SoftMax function, is where all neurons are the connected reason for using this design for training the model [50].

The model's accuracy on the tested photographs was about 74%. The model attempts to classify all items based on antecedent experience, however, this error persists because the dataset used to train the model only contains three distinct classes (apple, orange, and banana). If the "nonfruit" class had been included, the model might have been able to classify objects more accurately. The program uses the quicker R–CNN ResNet50 V1 640 × 640 model for real-time item detection. In this model, the R–CNN approach is applied rather than ResNet50, which has a slower architecture (Fig. 3.4). The COCO 2017 dataset was used to train the model, which can recognize up to 90 distinct object types. R-CNNs (region-based convolutional neural networks) are a well-known family of algorithms for their accuracy in object detection, despite the tedious inference process. Compared to earlier algorithms, R–CNN is a lot quicker (R–CNN and fast R–CNN). Alternatively, it does not use the selective search approach [51].

3.3 Recognition of human activity

An AI study attempted to develop a technique that could detect crucial human motions in real-time, including eating, which is vital in an Alzheimer's disease (AD) scenario for both patients and their caregivers. The study's main claim was that a system with a forecasting accuracy of at least 70% was achievable. The specific computer vision field known as activity recognition (events) involves identifying activities or events in videos [52]. The model was trained to recognize the activity in the video using the "Kinetics 400" dataset, which contains 400 human activities with at least 400 videos each [53]. Only the 400 behaviors that involved eating or drinking, such as "eating burgers" and "drinking," were used for the current study (Table 3.2).

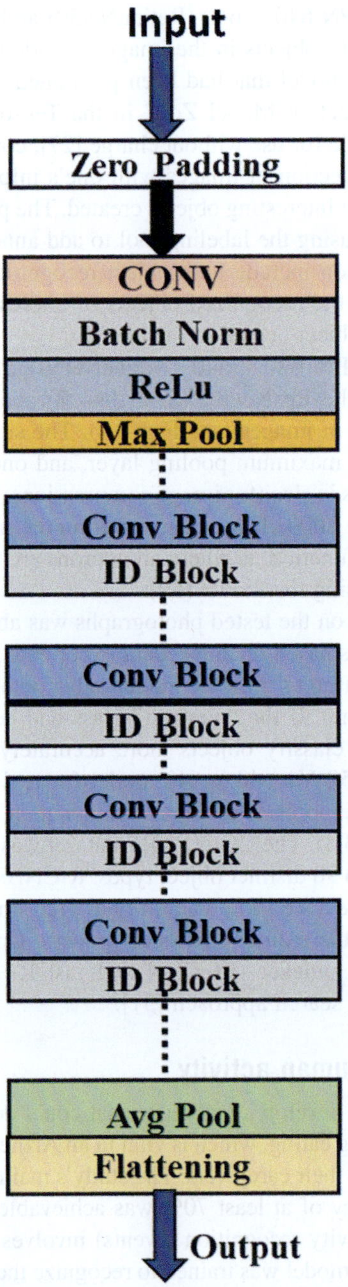

FIGURE 3.4

ResNet50 architecture.

Table 3.2 Machine learning methods.

Classification	Drug repurposing	Progression prediction	Biomarker identification
Deep convolutional neural network	Graph representing learning	Recurrent neural network	Association rule mining
Support vector machine	Network-based model	Multikernel learning	Naïve Bayes
Random forest	Recurrent neural network	Support vector machine	Association rule mining
Ensemble models	Graph neural network	Random forest	Deep neural network
Inception neural network	Random forest	Cox proportional hazard model	Differential analysis

In 2D convolutional networks, which are commonly used for image processing, have their convolutional kernel replaced by a 3D kernel. When trained on large datasets such as ImageNet for photographs, deep network architectures for image classification like ResNet, can achieve high accuracy. However, researchers have found that video classification can be accomplished with an appropriate video dataset [54,55]. The model mentioned in Ref. [54], when tested on the Kinetics 400 dataset, achieved an accuracy of 78.4%. It has been demonstrated that the classification of human behaviors in real-time is possible using 2D neural networks pretrained on the ImageNet dataset for image classification and 3D neural networks pretrained on Kinetics 400 for video classification with excellent results (see Fig. 3.5).

In [56], The inflating of 2D convolutional networks is the basis of a brand-new two-stream inflated 3D ConvNet architecture that is shown. Through the transformation of the deep network's 2D core for image classification into a 3D core, this design makes it possible to process the spatial and temporal components of films. To detect human activity in videos, this model was put to the test (see Fig. 3.6).

The Kinetics 400 test set [56] determined that this model's accuracy was 74.2%. On the Kinetics 400 test set, the first model—which utilized 3D spatial-temporal convolutional neural networks—performed better than the second model in terms of computation speed and accuracy. However, the first model fared better than the second in terms of recognizing all human activity in videos.

3.4 System architecture

The above-mentioned image categorization and human activity detection deep learning techniques can be successfully incorporated into a live, intelligent system created to help those with Alzheimer's (Fig. 3.7).

We developed an intelligent mobile/portable device that is specifically designed to assist individuals who are displaying early signs of Alzheimer's disease using the

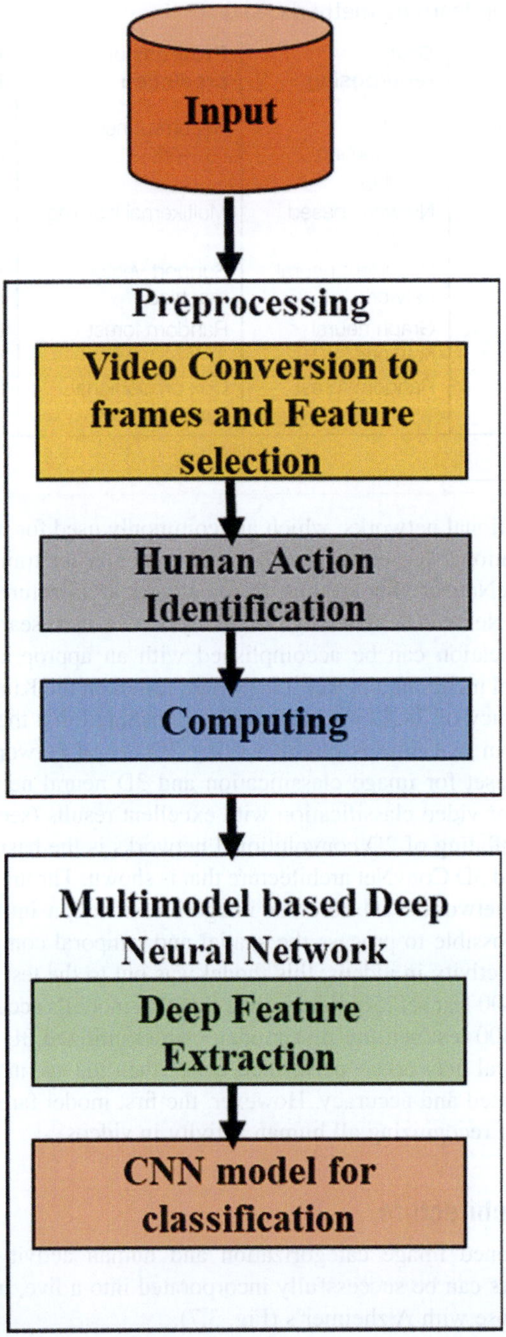

FIGURE 3.5

Classification of human actions in videos.

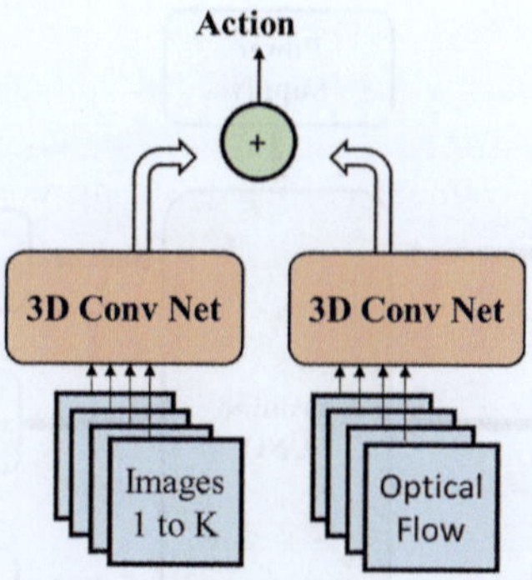

FIGURE 3.6
Two-stream 3D architecture.

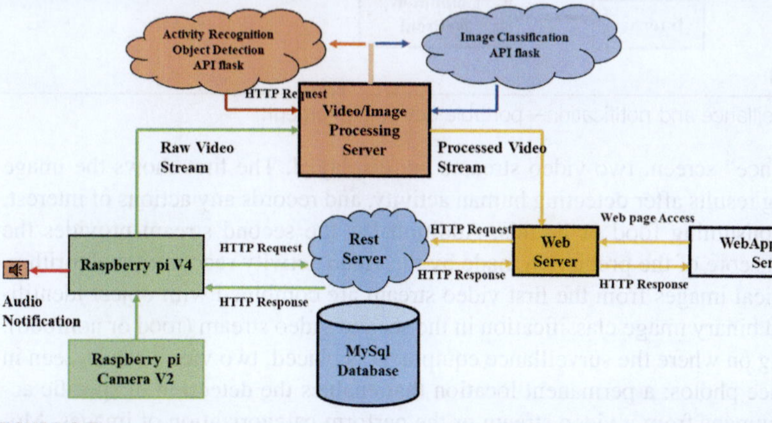

FIGURE 3.7
System diagram.

previously mentioned architecture as a foundation. Fig. 3.7 illustrates the main components of the system which include: a web application for patients and caregivers, (1) deep learning models accessed through a Python Flask interface, (2) a video monitoring application that connects with the deep learning models and the database, (3) a system database exposed similarly through REST services, and (4) the system database. This device uses a unique type of device, an audio notification, and video surveillance system.

3.5 Monitoring application

As shown in Fig. 3.8, a web application was created using modern methodologies to access the platform's capabilities. When the app is utilized, a webpage is launched that enables the carer to keep track of the Alzheimer's patient in real-time. On the

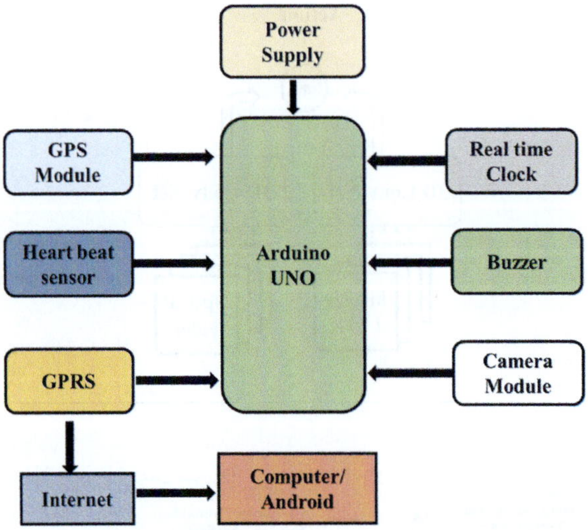

FIGURE 3.8

Video surveillance and notification—portable device equipment.

"surveillance" screen, two video streams are displayed. The first shows the image processing results after detecting human activity, and records any actions of interest, such as consuming food or liquids. Additionally, the second stream provides the likelihood score of the prediction made by the video activity recognition algorithm. The identical images from the first video stream are combined with object identification and binary image classification in the second video stream (food or nonfood). Depending on where the surveillance equipment is placed, two views can be seen in surveillance photos: a permanent location that enables the detection of specific activity of humans from a video stream or the perform categorization of images. Mobile placement, where the person wearing the portable device is constantly photographed from the viewpoint of the object of the observation.

a) Schedule

- The carer can adjust the settings to comply with guidelines for the time interval between meals to prevent overfeeding and to ensure that the patient stays hydrated by drinking fluids. Using the timer, the carer can modify the following: the email address for receiving notifications of the patient's activity, the duration of the patient's history in seconds (ranging from 1 to 120 s), and the time interval between notifications reminding the patient to drink fluids. These reminders and notifications also help the patient to manage their health and schedule medical appointments and screenings.
- A change to the auditory instructions telling the patient to put an end to eating to prevent overeating Although it is preprogrammed to say, "You should stop eating," the carer can adjust it to say, "You should stop eating," or something else that is more suggestive for the individual they are looking for.

- A modification to the email informed the carer that to avoid overfeeding, the patient needs to cease eating. Here is currently full of Changes that were made to the email alerting the carer that the patient had skipped meals to avoid sublimation. This is worth the predetermined amount. It can be "the patient forgot to eat," but it can also be customized to match the preferences of the carer.

b) Detection

The most efficient method for activity recognition is the use of neural networks, which are able to identify notable occurrences such as eating and drinking. The program is adaptable and can be accessed from any internet-connected device, regardless of the platform on which it is implemented.

c) Notifications

➢ The monitoring device, which is connected to a speaker and headphones, will play audio alerts for the individual with Alzheimer's disease that this system is monitoring. To maintain a balanced eating pattern, the messages they get to serve as a reminder of when to eat and drink:

➢ This kind of message is sent on determining when the patient is awake or has not eaten in the recommended amount of time—which the carer might have selected in the application—between meals. By default, the notification is of the type "You should eat" however, the carer may provide any customized but oblique form instead. Set the interval between reminder notifications so that if the patient does not eat for a time, The reminder to eat is given to you once again via an internet tool.

➢ Notification of forgotten hydration is delivered when an Alzheimer's patient is awake and has skipped more meals than usual or has gone longer than the caretaker-specified amount of time without drinking. The notification, which by default takes the form "You should drink," might also take any other specific but oblique form that the carer specifies. The notification is issued again if the patient does not drink within the window of time between notifications set in the web application.

➢ Realizing that he had neglected to eat or had neglected to If it is determined that the individual is eating too frequently or excessively, this form of signal is delivered even when they are following their sleep cycle because there is no chance of waking them up (the time specified by the guardian on the application). On identifying that the consumer consumes more food each day or too often because they are finishing their current meal faster than the necessary minimum gap between meals, the message is issued. Additionally, if a meal lasts longer than the advised number of minutes for eating, this notification is shown. The signal flashes at an interval every so often is the interval in seconds between database checks set up by the carer in the app, indicating that the individual is overeating. You should stop eating in this case, is the default message; however, the carer can set additional suggestive messages in the online program.

4. Discussions

No matter the network's objective—categorization of images, object recognition, or activity analysis—training a neural network takes a long time. It can take many days, depending on the hardware of the device. Following network training, the model that results can be applied with lightning-fast reaction times (2 s). Upon review of the performance score, confusion matrix, and inappropriately classified images, and creating a forecast for an image that is not in the dataset, it is possible to draw a few conclusions. Due to its excellent accuracy and low error rate in real-world use, the VGG16 model's architecture serves as the basis for The binary classification model's architecture and the ability to transfer knowledge from it makes it suitable for a variety of applications.

Considering how complicated the problem is and how much artificial vision is needed right now, the approach of object recognition in images is quite interesting. Object detection models are very effective because they make use of sophisticated deep learning networks and transfer learning algorithms to recognize things in almost real time. Artificial vision has a difficult time identifying human activities in videos. Both the timing of the video and picture processing is involved (Image order). To determine whether a near-real-time reaction speed is desirable, these models also require a lot of processing resources. This article's deep learning AI platform aims to give Alzheimer's patients back their freedom. but by reducing their burden on others and their carers' difficulties, too. Because of this, the Alzheimer's patient may be watched on camera, and the device can recognize when they are being fed or given water thanks to artificial intelligence.

5. Conclusion

The technique is designed to remind a person suffering from Alzheimer's to make healthy lifestyle decisions to prevent malnourishment, overeating, or dehydration. Since there is now no monitoring system that can recognize the actions of Alzheimer's disease patients and help them lead independent lives, there is a clear need for such a system. The program differs from other similar applications due to the novel It offers the concept of detecting human movement to assist those with dementia or Alzheimer's and others who care for them. The system is made to be able to detect activity from a variety of angles, including those where the surveillance equipment is portable, and the patient is carrying it, as well as those where it is a stationary sizable dataset that is focused on the desired behaviors for model training can be created to improve such a system. The absence of a suitable identifying mechanism, such as biometric identification using the fingerprint, face, voice, or iris, is a significant future implementation (and a current constraint) (thus reducing the urgent chance that the patient would forget his authentication credentials). The addition of data from recorded activities will be very beneficial to the physician when developing a diet for the patient based on their habits. The system should be able to distinguish between the things they eat to introduce a preestablished diet (from the output of algorithms for object detection and image categorization).

6. Future scope

Future developments should make the portable assistant device lighter and smaller by utilizing development boards and miniature batteries (for example, Raspberry Pi Zero). Additionally, a prototype of the system might be made to work with the cameras and computing power of today's handsets. The software program may stream visuals giving the user content while they are either food or drinking in addition to the warning or attention signs. To identify when different doors are opened, basic sensors can be incorporated into the overall architecture.

References

[1] Knopman DS, Amièva H, Petersen RC, Chételat G, Holtzman DM, Hyman BT, et al. Alzheimer disease. Nature Reviews Disease Primers 2021;7:1−21.

[2] Dementia Statistics|Alzheimer's Disease International (ADI). Available online: https://www.alzint.org/about/dementia-factsfigures/dementia-statistics/. [Accessed 28 December 2022].

[3] Biggs S, Carr A, Haapala I. Dementia as a source of social disadvantage and exclusion. Australasian Journal on Ageing 2019;38:26−33.

[4] Vally M, Kathrada F. Understanding Alzheimer's disease. In: South African family practice; 2019.

[5] What you need to know about dementia and eating disorders—the Alzheimer's site news. Available online: https://blog.thealzheimerssite.greatergood.com/eating-disorders/. [Accessed 26 December 2022].

[6] Wilson K, Dewing J. Strategies to prevent dehydration in older people with dementia: a literature review. Nursing Older People 2019;32:27−33.

[7] Aljojo N, Alotaibi RM, Alharbi B, Alshutayri A, Tariq Jamal A, Banjar A, et al. Alzheimer assistant: a mobile application using Machine Learning. Revista Română de Informatică și Automatică 2020;30:7−26.

[8] Dementia: technology to stay independent—SCIE. Available online: https://www.scie.org.uk/dementia/support/technology/stay-independent. [Accessed 26 December 2022].

[9] Sharma S, Dudeja RK, Aujla GS, Bali RS, Kumar N. DeTrAs: deep learning-based healthcare framework for IoT-based assistance of Alzheimer patients. Neural Computing & Applications 2020.

[10] Roopaei M, Rad P, Prevost JJ. A wearable IoT with complex artificial perception embedding for Alzheimer patients. In: 2018 world automation congress (WAC); 2018. p. 1−6.

[11] Gollapudi S. Learn computer vision using OpenCV: with deep learning CNNs and RNNs. 2019.

[12] SriPreethaa K R, Yuvaraj N, Jenifa G. Ensemble machine learning approach for diabetes prediction. 2020.
[13] Boehle MD, Eitel F, Weygandt M, Ritter K. Layer-wise relevance propagation for explaining deep neural network decisions in MRI-based Alzheimer's disease classification. Frontiers in Aging Neuroscience 2019;11:194.
[14] Nguyen M, He T, An L, Alexander DC, Feng J, Yeo BT. Predicting Alzheimer's disease progression using deep recurrent neural networks. bioRxiv; 2019.
[15] Shi B, Chen Y, Zhang P, Smith CD, Liu J. Nonlinear feature transformation and deep fusion for Alzheimer's Disease staging analysis. Pattern Recognit 2015;63:487–98.
[16] Faturrahman M, Wasito I, Hanifah N, Mufidah R. Structural MRI classification for Alzheimer's disease detection using deep belief network. 2017 11th International Conference on Information & Communication Technology and System (ICTS). 2017. p. 37–42.
[17] Nawaz H, Maqsood M, Afzal S, Aadil F, Mehmood I, Rho S. A deep feature-based real-time system for Alzheimer disease stage detection. Multimedia Tools and Applications 2020:1–19.
[18] Liang S, Gu Y. Computer-aided diagnosis of Alzheimer's disease through weak supervision deep learning framework with attention mechanism. Sensors 2020;21:220.
[19] Katabathula S, Wang Q, Xu R. Predict Alzheimer's disease using hippocampus MRI data: a lightweight 3D deep convolutional network model with visual and global shape representations. Alzheimer's Research & Therapy 2021;13:104.
[20] Gao S, Lima D. A review of the application of deep learning in the detection of Alzheimer's disease. International Journal of Cognitive Computing in Engineering 2021; 3:1–8.
[21] Yuvaraj N, SriPreethaa KR. Diabetes prediction in healthcare systems using machine learning algorithms on Hadoop cluster. Cluster Computing 2017:1–9.
[22] Saleem TJ, Zahra SR, Wu F, Alwakeel AS, Alwakeel MM, Jeribi F, et al. Deep learning-based diagnosis of Alzheimer's disease. Journal of Personalized Medicine 2022;12:815.
[23] Chien Y, Hong S, Cheah W, Yao L, Chang Y, Fu L. An automatic assessment system for Alzheimer's disease based on speech using feature sequence generator and recurrent neural network. Scientific Reports 2019;9:1–10.
[24] Roshanzamir A, Aghajan HK, Soleymani Baghshah M. Transformer-based deep neural network language models for Alzheimer's disease risk assessment from targeted speech. BMC Medical Informatics and Decision Making 2021;21:92.
[25] Allegra D, Battiato S, Ortis A, Urso S, Polosa R. A review on food recognition technology for health applications. Health Psychology Research 2020;8:9297.
[26] Munkevik P, Hall G, Duckett T. A computer vision system for appearance-based descriptive sensory evaluation of meals. Journal of Food Engineering 2007;78:246–56.
[27] Sivaranjani R, Yuvaraj N. Earlier prediction of cardiovascular disease using IoT and deep learning approaches. In: Advances in computerized analysis in clinical and medical imaging; 2019.
[28] Kong F, Tan J. DietCam: automatic dietary assessment with mobile camera phones. Pervasive and Mobile Computing 2012;8:147–63.
[29] Lu Y, Stathopoulou T, Vasiloglou MF, Christodoulidis S, Blum B, Walser T, et al. An artificial intelligence-based system for nutrient intake assessment of hospitalised patients*. 2019 41st Annual International Conference of the IEEE Engineering in Medicine and Biology Society (EMBC). 2019. p. 5696–9.

[30] Jo T, Nho K, Saykin AJ. Deep learning in Alzheimer's disease: diagnostic classification and prognostic prediction using neuroimaging data. Frontiers in Aging Neuroscience 2019;11:220.

[31] Jack CR, Bennett DA, Blennow K, Carrillo MC, Dunn B, Haeberlein SL, et al. NIA-AA research framework: toward a biological definition of Alzheimer's disease. Alzheimer's and Dementia: The Journal of the Alzheimer's Association 2018;14:535–62.

[32] Vyas A, Mehta MA, Kotecha K, Pandya S, Alazab M, Gadekallu TR. Tear film breakup time-based dry eye disease detection using convolutional neural network. Neural Computing & Applications 2022.

[33] Shymanska A. Book review of "AI superpowers: China, Silicon valley, and the new world order" by Kai-Fu Lee. 2019.

[34] Sharma DK, Chatterjee M, Kaur G, Vavilala S. Deep learning applications for disease diagnosis. In: Deep learning for medical applications with unique data; 2022.

[35] De Fauw J, Ledsam JR, Romera-Paredes B, Nikolov S, Tomašev N, Blackwell S, et al. Clinically applicable deep learning for diagnosis and referral in retinal disease. Nature Medicine 2018;24:1342–50.

[36] Storås AM, Strümke I, Riegler M, Grauslund J, Hammer HL, Yazidi A, et al. Artificial intelligence in dry eye disease. Ocular Surface 2021;23:74–86.

[37] Le Berre C, Sandborn WJ, Aridhi S, Devignes M, Fournier L, Smaïl-Tabbone M, et al. Application of artificial intelligence to gastroenterology and hepatology. Gastroenterology 2019;158:76–94.e2.

[38] Thrall J, Li X, Li Q, Cruz C, Do S, Dreyer KJ, et al. Artificial intelligence and machine learning in radiology: opportunities, challenges, pitfalls, and criteria for success. Journal of the American College of Radiology: JACR 2018;15(3 Pt B):504–8.

[39] Yeh C, Yu SX, Lin MC. Meibography phenotyping and classification from unsupervised discriminative feature learning. Translational Vision Science & Technology 2021;10:4.

[40] Ramana K, Kumar MR, Sreenivasulu K, Gadekallu TR, Bhatia S, Agarwal P, et al. Early prediction of lung cancers using deep saliency capsule and pre-trained deep learning frameworks. Frontiers in Oncology 2022;12:886739.

[41] Ganesan J, Inbarani HH. Soft set based feature selection approach for lung cancer images. ArXiv, abs/1212.5391. 2012.

[42] IMDB-WIKI—500k+ face images with age and gender labels. Available online: https://data.vision.ee.ethz.ch/cvl/rrothe/imdb-wiki/. [Accessed 26 December 2022].

[43] Rothe R, Timofte R, Gool LV. Deep expectation of real and apparent age from a single image without facial landmarks. International Journal of Computer Vision 2018;126: 144–57.

[44] Rothe R, Timofte R, Gool LV. DEX: deep expectation of apparent age from a single image. 2015 IEEE International Conference on Computer Vision Workshop (ICCVW). 2015. p. 252–7.

[45] Kaggle. Furniture detector. Available online: https://www.kaggle.com/datasets/akkithetechie/furniture-detector. [Accessed 26 December 2022].

[46] Kaggle. Food images (Food-101). Available online: https://www.kaggle.com/datasets/kmader/food41. [Accessed 28 December 2022].

[47] Jadon S. SSM-net for plants disease identification in low data regime. 2020 IEEE/ITU International Conference on Artificial Intelligence for Good (AI4G). 2020. p. 158–63.

[48] Classification: True vs. False and Positive vs. Negative|Machine Learning|Google Developers. Available online: https://developers.google.com/machine-learning/crash-course/classification/true-false-positive-negative?hl=en. [Accessed 26 December 2022].

[49] GitHub. Heartexlabs/LabelImg: LabelImg is a graphical image annotation tool and label object bounding boxes in images. Available online: https://github.com/heartexlabs/labelImg. [Accessed 26 December 2022].

[50] Peng J, Kang S, Ning Z, Deng H, Shen J, Xu Y, et al. Residual convolutional neural network for predicting response of transarterial chemoembolization in hepatocellular carcinoma from CT imaging. European Radiology 2019;30:413−24.

[51] Liu Y, Ma Z, Liu X, Ma S, Ren K. Privacy-preserving object detection for medical images with faster R-CNN. IEEE Transactions on Information Forensics and Security 2019;17:69−84.

[52] Activity recognition—Wikipedia. Available online: https://en.wikipedia.org/wiki/Activity_recognition. [Accessed 26 December 2022].

[53] Kay W, Carreira J, Simonyan K, Zhang B, Hillier C, Vijayanarasimhan S, et al. The kinetics human action video dataset. ArXiv; 2017. abs/1705.06950.

[54] Hara K, Kataoka H, Satoh Y. Can spatiotemporal 3D CNNs retrace the history of 2D CNNs and ImageNet?. 2018 IEEE/CVF Conference on Computer Vision and Pattern Recognition. 2017. p. 6546−55.

[55] Rosebrock A. Human activity recognition with OpenCV and deep learning. Philadelphia, PA, USA: PyImageSearch; 2019.

[56] Carreira J, Zisserman A. Quo Vadis, action recognition? A new model and the Kinetics dataset. 2017 IEEE Conference on Computer Vision and Pattern Recognition (CVPR). 2017. p. 4724−33.

CHAPTER

Rehabilitation for individuals with autism spectrum disorder using mixed reality virtual assistants

4

A. Jenefa[1], A. Lincy[2] and J. Anitha[3]

[1]Department of Computer Science and Engineering, Karunya Institute of Technology and Sciences, Coimbatore, Tamil Nadu, India; [2]Department of Computer Science and Engineering, National Engineering College, Kovilpatti, Tamil Nadu, India; [3]Department of Electronics and Communication Engineering, Karunya Institute of Technology and Sciences, Coimbatore, Tamil Nadu, India

1. Introduction

Medical expenses for autism spectrum disorder (ASD), a neurodevelopmental disorder with high indirect costs, can be greatly reduced with early treatment. Inconveniently, the diagnostic process for autism spectrum disorder takes longer and is expensive. As the global incidence of ASD continues to rise, there is an immediate requirement for the advancement of accessible and reliable screening tools. Healthcare providers and individuals will greatly benefit from a quick and easy screening for autism spectrum disorder to determine whether or not a further clinical evaluation is recommended. The term ASD refers to a group of neurodevelopment disorders characterized by difficulties in social interaction, information exchange, and behavior, but may not portray a formal diagnostic category. The prevalence of autism spectrum disorder among children worldwide was recently estimated to be 1 in 100, according to a study released in Autism Research (ASD). It is up from the 2012 worldwide prevalence report, which estimated that 62 out of every 10,000 kids were autistic. Mixed reality (MR) is a technology that combines the real and virtual worlds, and it has been increasingly used in the field of autism to improve various aspects of the lives of individuals with autism. Some examples of MR-based technologies that have been developed for autism include:

Virtual reality (VR) and augmented reality (AR) systems can be used to provide social and communication training to individuals with autism. These systems can help individuals practice social skills in a controlled environment and receive feedback on their performance. MR games and activities can be used to motivate and engage individuals with autism in learning and therapy sessions. These MR

Computational Intelligence and Deep Learning Methods for Neuro-rehabilitation Applications
https://doi.org/10.1016/B978-0-443-13772-3.00011-X

experiences can be tailored to the interests and needs of the individual and can be used to teach a wide range of skills, including social skills, communication skills, and academic skills. MR-based assessment and evaluation tools can be used to measure the progress and performance of individuals with autism. These tools can provide a more accurate and objective assessment of an individual's abilities and can be used to track their progress over time. MR-based assistive technology can help individuals with autism perform everyday tasks and activities more independently. These MR tools can provide step-by-step instructions, visual prompts, and other forms of support to help individuals with autism complete tasks more effectively.

In recent times, modern technologies like MR have become more and more important in the treatment of several disorders. This is because these devices can provide users with content that is adaptable to their needs and that they can keep a familiar environment while they are undergoing therapy. Although there is no unified description of MR, the technique is typically described as the merging of the real and virtual worlds. This is because they can offer personalized experiences based on the needs of the users and keep a familiar environment during therapy. For people with ASD, the mixed reality—virtual assistants is a quick way to learn something therapeutic at home. It helps them make eye contact with their caregivers and connect different facial expressions with different emotions. Fig. 4.1 represents the graphical representation of virtual reality, augmented reality, and mixed reality.

MR app HoloLearn uses Microsoft HoloLens to aid those with cognitive disabilities in becoming more independent in their daily lives. HoloLearn is a MR platform that places the user in an immersive environment based on their immediate physical surroundings, allowing them to learn basic, everyday tasks interestingly and interactively, with the aid of a virtual assistant if necessary. HoloLens is a mixed reality device developed by Microsoft. It is a self-contained, holographic computer that

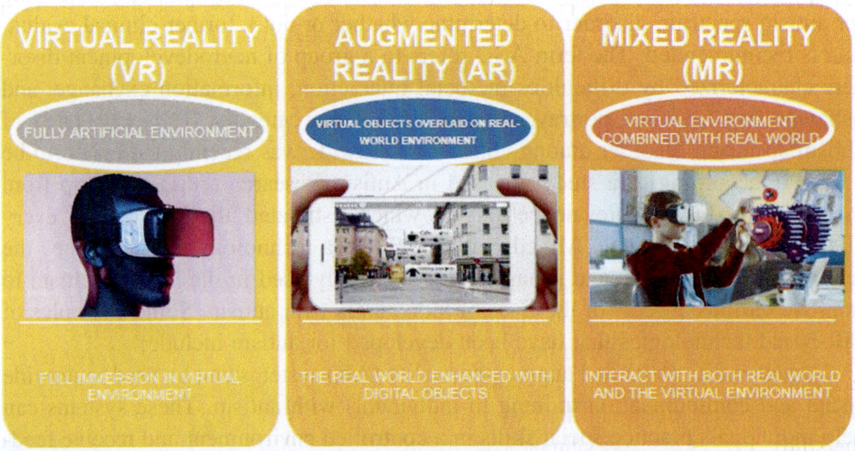

FIGURE 4.1

A graphical representation of virtual reality, augmented reality, and mixed reality.

enables users to interact with digital content and information in a natural, immersive way. HoloLens uses a transparent display to superimpose 3D holograms over the real world, allowing users to see and interact with both physical and virtual objects at the same time. HoloLearn is a mixed reality app that is designed to work with the HoloLens, and it may allow users to learn and explore new concepts and information through interactive holographic experiences. The HoloLens is a headset that allows users to see and interact with digital content in the real world. It uses a combination of AR and VR to create a mixed reality environment in which digital content is integrated into the physical world. The HoloLearn app may use the HoloLens to create interactive, immersive learning experiences that combine the real world with digital elements.

This article discusses the use of MR in the diagnosis and rehabilitation of individuals with ASD. This paper's outline is as follows. In Section 2, we talk about the state-of-the-art methods that have been proposed for accurate ASD detection In Section 3, we will discuss the MR used for automatic ASD detection using HoloLearn. Section 4 covers the introduction of MR-based rehabilitation tools for helping those with ASD. In Section 5, we look at the difficulties that arise in using MR for ASD diagnosis and Future endeavors, and the paper summarizes in Section 6.

2. Related work

Recent years have seen a rapid spread of MR-based technologies, which are now used in a variety of fields, from building design and construction to speed up the urban development process to training and learning tools for doctors in the medical field. In the field of education, there are some examples of technologies made for kids, like the one described in [1], which is an MR system meant to help kids learn better by working together, piquing their curiosity, and teaching them to work together. Mixed reality technologies, which include AR and VR, have shown promise as tools to assist in the treatment of autism. These technologies can be used to create immersive and engaging learning environments that can be customized to the needs of individuals with autism. Here are a few examples of mixed reality technologies that have been used to assist in the treatment of autism, along with references to relevant studies:

In the study, Passarello et al. [2] developed a mixed reality play environment for children with autism and evaluated its effectiveness in promoting social interaction and play behavior. In the study, Mills et al. [3] examined the use of VR in the treatment of sensory processing problems in individuals with autism. In the study, Sadeghi Milani et al. [4] explored the use of technology, including VR, to facilitate play in children with autism. In a pilot study, Ke et al. [5] investigated the use of VR for social skills training in children with high functioning autism spectrum disorder. In the study, Lee et al. [6] evaluated the use of VR in the treatment of children with autism. In the study, Zhang et al. [7] examined the use of VR in the treatment of anxiety in children with autism. In a meta-analysis, Alahmari et al. [8] analyzed the effectiveness of VR exposure therapy for anxiety and related disorders. In the study, Farashi et al. [9] explored the use of VR exposure therapy for social phobia in nonverbal patients with autistic disorders. In the study, Liu et al. [1] reviewed the use of VR as a treatment for anxiety disorders in children and adolescents. In the study, Merlo et al. [10] examined the use of VR exposure therapy for anxiety and related disorders.

Virtual reality exposure therapy (VRET): VRET is a type of cognitive behavioral therapy that uses VR to expose individuals with anxiety disorders, including those with autism, to feared or avoided situations in a controlled and safe environment [11]. Augmented reality social skills training: AR can be used to create interactive and engaging social skills training programs for individuals with autism [12].Virtual reality functional communication training: VR can be used to teach functional communication skills to individuals with autism who have limited verbal abilities [13]. A review by Ref. [14] entitled "Augmented and virtual reality in autism spectrum disorder: A systematic review" found that mixed reality technologies can be effective in improving social interaction, communication, and behavioral problems in individuals with autism. A study by Ref. [15] entitled "Virtual reality social cognition training for individuals with high-functioning autism: A randomized controlled trial" found that VR social cognition training was effective in improving social skills in individuals with high-functioning autism. A study by Kim et al. [16] entitled "The effects of virtual reality-based social skills training on the social interaction of children with autism spectrum disorder" found that VR-based social skill training was effective in improving social interaction in children with autism. Table 4.1 shows the techniques used for rehabilitation for individuals with an autism spectrum disorder.

Overall, the studies suggest that MR can be effective in improving social skills and reducing anxiety in individuals with autism. The majority of the studies are case studies with small sample sizes, although there is also a meta-analysis and a pilot study with a larger sample size. While the findings are promising, further research with larger sample sizes and more controlled designs is needed to fully understand the effectiveness of VR in treating symptoms of autism.

2.1 Common interventions in rehabilitation of ASD

Communication and social interaction can be negatively impacted by ASD, a neurodevelopmental disease. It is a disorder that lasts a lifetime and is usually identified in children; however, adults can also be affected. Interventions used in ASD patients' rehabilitative care typically target the patients' language and social functioning as well as their behavior. Examples of such measures could be as follows:

Speech therapy: People with ASD can benefit greatly from speech therapy, which focuses on both verbal and nonverbal communication skills development. Individuals with ASD can benefit from this sort of therapy by strengthening their communication abilities, such as their capacity to

Table 4.1 Techniques used for rehabilitation for individuals with autism spectrum disorder.

Authors	Study design	Sample size	Main findings
Passarello et al. [2]	Case study	1	The mixed reality play environment was effective in engaging and motivating the participant, and in promoting social interactions and communication.
Mills et al. [3]	Case study	1	VR was effective in reducing sensory processing problems and improving social interactions in the participant.
Sadeghi Milani et al. [4]	Case study	1	The technology was effective in engaging and motivating the participant, and in promoting social interactions and communication.
Ke et al. [5]	Pilot study	8	VR was effective in improving social skills in the participants.
Lee et al. [6]	Case study	1	VR was effective in reducing anxiety and improving social skills in the participant.
Zhang et al. [7]	Case study	1	VR was effective in reducing anxiety and improving social skills in the participant.
Alahmari et al. [8]	Meta-analysis	N/A	VR was effective in reducing anxiety in individuals with anxiety disorders.
Farashi et al. [9]	Case study	1	VR was effective in reducing social phobia in the participant.
Shahmoradi et al. [17]	Randomized controlled trial	75 children with ASD	Mixed reality program improved social communication skills in children with ASD.
Howlin et al. [18]	Randomized controlled trial	48 children with ASD	Mixed reality program improved social communication skills in children with ASD.
Kaale et al. [19]	Randomized controlled trial	50 children with ASD	Mixed reality program improved social communication skills in children with ASD.
Rossignol et al. [20]	Systematic review and meta-analysis	N/A	Mixed reality interventions significantly improved social skills in children with ASD.
Philip et al. [21]	Systematic review	N/A	Virtual and augmented reality interventions can be effective for children with ASD.

comprehend and use language. Individuals with ASD can benefit greatly from speech and language therapy in terms of developing and strengthening their communication and language abilities. As a result, it is possible that both speaking and listening skills will need to be developed (understanding).

3.2 HoloLearn: Unpacking the mechanics of mixed reality learning

HoloLearn is a mixed reality learning platform that utilizes the Microsoft HoloLens headset to provide immersive, interactive learning experiences for person with autism. The HoloLens headset allows to see and interact with virtual objects and environments in their physical space, creating a more engaging and interactive learning experience. The HoloLens headset is equipped with a range of sensors and cameras that allow it to detect and track the user's movements and surroundings. It also has a built-in microphone and speakers, allowing the user to communicate with the system and hear virtual objects and sounds. HoloLearn uses a variety of software programs and applications to create interactive learning experiences for students with autism. These may include virtual reality simulations, augmented reality games, and interactive educational videos and activities. Students with autism can use the HoloLens headset to learn a wide range of skills and concepts, including social interaction, communication, problem-solving, and critical thinking. The immersive and interactive nature of the HoloLens allows students to more fully engage with the learning material and apply their newfound skills in real-world situations. Overall, HoloLearn is a powerful tool for improving the education and learning outcomes of students with autism, using the latest in mixed reality technology to create engaging and interactive learning experiences. So, the diagram of HoloLearn using Microsoft HoloLens can be represented in Fig. 4.2 as follows:

3.3 Exploring the world of autism: A HoloLearn activity

HoloLearn is a virtual reality platform that provides educational and therapeutic activities for individuals with autism.

3.3.1 Setting the table activity

This activity involves setting the table for a meal, including placing the plates, utensils, and other items in their proper places. This helps children with autism practice

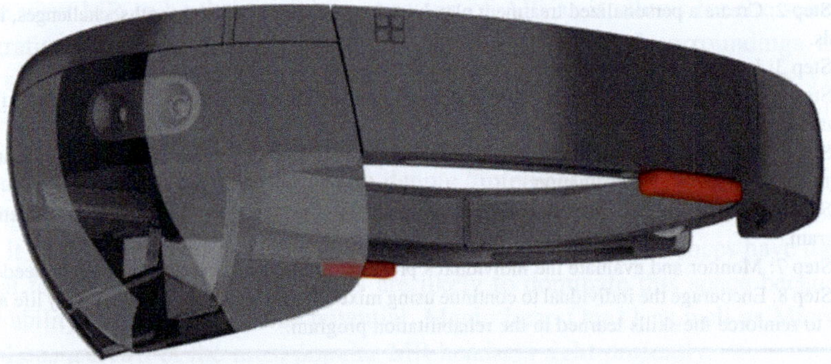

FIGURE 4.2

Diagram of HoloLearn being used by Microsoft HoloLens.

following instructions and understanding spatial concepts. To begin the lay-the-table activity for autism in HoloLearn using Microsoft HoloLens, the user would need to put on the HoloLens headset and follow the prompts on the screen.

Step 1: The user would be shown a virtual table in their field of vision, with all of the necessary items for setting the table displayed around the room.
Step 2: The user would be given verbal instructions on how to properly lay the table, such as "place the placemats on the table" or "set the forks on the left side of the plate."
Step 3: The user would then physically walk around the room and interact with the virtual objects to complete the task. For example, they would reach out and grab a placemat from the virtual shelf and place it on the virtual table.
Step 4: As the user completes each step, they would receive verbal feedback and encouragement, such as "good job" or "well done."
Step 5: Once the user has completed all of the steps, they would be given a final review of the table to ensure everything is in its proper place. If any mistakes are made, the user would be shown where the mistakes are and given the opportunity to correct them.

Overall, this activity helps individuals with autism practice important life skills such as following instructions, problem-solving, and spatial awareness in a virtual environment. The use of the HoloLens headset allows for a more immersive and interactive experience, making it easier for the user to understand and complete the task.

3.3.2 HoloLearn's garbage collection efforts

This activity involves sorting through different objects and determining which ones are garbage and which ones are recyclables or items that can be reused. This helps children with autism improve their sorting and categorization skills. One activity that is offered on the platform is a garbage-sorting game. This activity involves the user sorting different types of virtual garbage into the correct recycling bins. This activity helps individuals with autism improve their fine motor skills, problem-solving skills, and attention to detail. It also helps teach them about the importance of recycling and taking care of the environment. Overall, the garbage sorting game is a fun and engaging way for individuals with autism to learn and practice important skills.

3.3.3 HoloLearn memory activity

This activity involves using memory techniques to remember lists of items or sequences of actions. This helps children with autism improve their short-term memory and attention to detail. In this memory activity, participants will use the HoloLearn platform to engage in a series of challenges designed to improve their memory skills. The activity will begin with a brief introduction to the importance of memory in everyday life and the ways in which it can be improved through practice and training. Participants will then be presented with a series of memory challenges, including visual and verbal tasks, that will require them to remember specific details and patterns. These challenges will be presented in a variety of formats, including interactive games, virtual reality simulations, and educational videos. Throughout the activity, participants will have the opportunity to track their progress and receive feedback on their performance. They will also be able to compete with other participants and see how they stack up against their peers. Overall, this memory activity will provide a fun and engaging way for participants to improve their memory skills and better understand the inner workings of their minds.

3.3.4 Virtual assistant

This activity involves using a virtual assistant, such as Amazon's Alexa or Google Assistant, to interact with and perform tasks. This helps children with autism improve their communication skills and familiarize themselves with technology. For example, a child with autism may use a virtual assistant to set reminders or alarms, ask for information about a specific topic, or play music or games. This allows the child to practice their communication skills by giving commands and asking questions and helps them understand how to use technology to accomplish tasks. Using a virtual assistant can also help children with autism improve their social skills by allowing them to engage in conversation with a computer-generated voice. This can be especially useful for children who may struggle with social interactions with other people. Additionally, virtual assistants can provide a sense of independence and autonomy for children with autism, as they can complete tasks and access information on their own. This can help boost their self-esteem and confidence. Overall, using a virtual assistant can be a fun and useful tool for children with autism to improve their communication and social skills, as well as their understanding of technology.

3.3.5 Dress-up activity

This activity involves selecting and dressing in different costumes or clothing items. This helps children with autism practice decision-making and self-expression. In the dress-up activity in HoloLearn rehabilitation for ASD, the therapist will work with the child to practice social skills and emotional regulation through imaginative play. The therapist will provide a variety of dress-up clothes and accessories, such as hats, scarves, and shoes, and will encourage the child to choose items to create a character or persona. The therapist will then guide the child through the process of putting on the clothes and accessories, encouraging them to use appropriate communication and social skills, such as asking for help or taking turns with the therapist. As the child plays with the dress-up clothes, the therapist will encourage them to express their thoughts and feelings about the character they are creating and to practice empathy and perspective-taking as they imagine how their character might feel or act in different situations. Throughout the activity, the therapist will provide positive reinforcement and support for the child's efforts and will work with them to identify and manage any challenges or emotions that may arise. The dress-up activity can be modified to fit the needs and abilities of each individual child and can be repeated with different themes or characters to continue building social skills and emotional regulation over time.

4. ASD pathways: Tools for rehabilitation and progress

There are several mixed reality-based rehabilitation tools that have been developed for individuals with autism. These tools use virtual or augmented reality technology to create interactive, immersive environments that can help individuals with autism improve social skills, communication, and other behaviors. Some examples of mixed reality-based rehabilitation tools for autism include the following:

4.1 Autism XR

This is a virtual reality tool that provides social and emotional learning experiences for individuals with autism. It includes a variety of interactive scenarios and games that can help individuals practice social skills and improve communication.

Autism XR is a VR technology designed specifically for individuals with autism. It is believed that VR can help individuals with autism to better understand and process social cues, practice social skills, and improve communication skills. It is also thought to be an effective tool for reducing anxiety and increasing self-regulation. Some examples of Autism XR programs include VR social skills training, VR exposure therapy, and VR relaxation techniques. It is important to note that while there is some evidence supporting the use of VR in autism treatment, more research is needed to fully understand its potential benefits.

4.2 Real social skills

This is an augmented reality tool that helps individuals with autism practice social skills in real-world situations. It uses interactive videos and games to teach individuals how to interact with others and navigate social situations. The Real Social Skills app is a rehabilitation tool specifically designed for individuals with autism. It aims to help individuals with autism improve their social skills, communication skills, and overall social interaction skills. The app features a variety of interactive activities and games that focus on different social skills, including eye contact, body language, conversation skills, and emotion recognition. It also includes educational videos and written materials that provide further information on social skills and how to effectively practice them in real-life situations.

One of the key features of the Real Social Skills app is its personalized progress-tracking system. It allows users to track their progress and see where they may need additional support or practice. It also provides feedback and guidance on how to improve social skills in different situations. The Real Social Skills app is a valuable tool for individuals with autism who are looking to improve their social skills and interact more confidently with others. It can be used in conjunction with therapy sessions or as a standalone tool for daily practice.

4.3 PRISMA

This is a virtual reality tool that provides immersive social learning experiences for individuals with autism. It includes a variety of interactive scenarios and games that can help individuals practice social skills, improve communication, and manage emotions. The PRISMA app is a rehabilitation tool designed specifically for individuals with autism. It uses a variety of interactive games and activities to help improve social skills, communication skills, and overall cognitive functioning. One of the key features of the app is its ability to track progress and adjust the difficulty level of activities based on the user's abilities. This allows for a personalized and tailored approach to rehabilitation, ensuring that the user is challenged but not overwhelmed. The app also includes a variety of tools and resources for parents and caregivers, including tips and strategies for interacting with and supporting individuals with autism.

Overall, mixed reality-based rehabilitation tools have the potential to be effective interventions for individuals with autism, as they can provide immersive and interactive learning experiences that can help individuals practice and improve important skills. Overall, all of these apps seem to have similar goals of teaching social skills and communication skills to individuals with autism through interactive methods. However, they differ in their approach and accessibility. Autism XR and HoloLearn both use virtual reality technology to create immersive learning experiences, but this may not be accessible to everyone due to the cost of VR technology. Prisma and Real Social Skills, on the other hand, are both accessible through a smartphone or tablet and do not require VR technology. In terms of accuracy, all of these apps may have some limitations in accurately representing real-world social situations. However, Real Social Skills and HoloLearn may have a higher accuracy due to their interactive and personalized learning methods. Ultimately, it may be best to try out a few different apps to see which one works best for an individual's needs and preferences. Table 4.2 represents the comparison of different App of mixed reality-based rehabilitation tools for autism.

4.4 Discussions

Fifty individuals with disabilities participated in two experimental studies using HoloLearn, one with a virtual assistant and one without. In the experimental setting with the virtual assistant, participants reported feeling more confident and capable of completing tasks and navigating the program. They appreciated the additional support and guidance provided by the assistant and felt that it helped them to better understand and retain information. In contrast, participants in the experimental setting without the virtual assistant reported feeling more frustrated and overwhelmed. They struggled to complete tasks on their own and found it difficult to navigate the program without guidance. Some participants even reported feeling discouraged and giving up on certain tasks because of the lack of support.

Table 4.2 Comparison of different App of mixed reality-based rehabilitation tools for autism.

App	Description	Pros	Cons	Accuracy
Autism XR	Virtual reality app that helps individuals with autism learn social skills and communication skills through interactive scenarios	- Allows for immersive learning experience - Can be used in a controlled environment - Can be customized to individual needs	- May not be accessible to everyone due to the cost of VR technology - May not accurately represent real-world social situations	Moderate
HoloLearn	Virtual reality app that helps individuals with autism learn social skills and communication skills through interactive scenarios	- Allows for immersive learning experience - Can be used in a controlled environment - Can be customized to individual needs	- May not be accessible to everyone due to cost of VR technology - Requires access to virtual reality technology	High
Prisma	App that teaches social skills through interactive games and activities	- Allows for personalized learning - Can be used anytime, anywhere	- May not be as immersive as virtual reality apps - May not accurately represent real-world social situations	Moderate
Real social skills	App that teaches social skills through interactive games, videos, and activities	- Allows for personalized learning - Can be used anytime, anywhere	- May not be as immersive as virtual reality apps - May not accurately represent real-world social situations	High

Overall, the results of the experiment showed that the presence of a virtual assistant significantly improved the experience and outcomes of individuals diagnosed with disabilities using HoloLearn. The assistant provided much-needed support and guidance, helping participants to feel more confident and capable of using the program. The results showed that the group using the virtual assistant had significantly higher scores on various assessments, including cognitive skills, communication abilities, and overall engagement with the HoloLearn program. They also reported higher levels of satisfaction and motivation and lower levels of frustration and boredom.

On the other hand, the group without a virtual assistant struggled more with navigating the program and understanding the material, leading to lower scores and higher levels of frustration. They also reported feeling less motivated and engaged, and more anxious and isolated. Overall, the results suggest that the virtual assistant plays a crucial role in supporting and empowering individuals with disabilities to learn and engage with new technologies. It provides a more personalized and accessible learning experience, allowing them to fully participate and benefit from the educational opportunities offered by HoloLearn. Table 4.3 represents the comparison of HoloLearn with other reality technologies.

Reality refers to the objective, physical world around us. In terms of accuracy, reality is high because it is based on the actual physical world. However, in terms of speed, reality may be slower because it requires physical movement and may be limited by the laws of physics. User experience, in reality, is generally good because it is familiar and easy to navigate. Overall, reality scores a 4 out of 5. Virtual reality refers to a simulated environment that is created using computer technology. In terms of accuracy, virtual reality can be high as long as it is designed to be realistic. However, in terms of speed, virtual reality may be slower because it requires the use of a computer or device to navigate. User experience in virtual reality can be good, but it may also feel somewhat disorienting or artificial compared to reality. Overall, virtual reality scores a 3.5 out of 5. Augmented reality refers to the overlay of digital information onto the physical world. In terms of accuracy, augmented reality can be medium to high depending on the quality of the information being overlaid. In terms of speed, augmented reality is generally excellent because it is real time and does not require a separate device or computer to access. User experience in augmented reality is generally excellent because it is interactive and can enhance the physical world. Overall, augmented reality scores a 4.5 out of 5. Mixed reality refers to the combination of virtual and physical elements in a single environment.

Table 4.3 Comparison of HoloLearn with other reality technologies.

Reality	Accuracy	Speed	User experience	Overall score
Reality	High	Medium	Good	4/5
Virtual reality	High	Low	Good	3.5/5
Augmented reality	Medium	High	Excellent	4.5/5
Mixed reality	High	High	Excellent	5/5

In terms of accuracy, mixed reality is generally high because it combines both real and virtual elements. In terms of speed, mixed reality is generally excellent because it is real-time and does not require a separate device or computer to access. User experience in mixed reality is generally excellent because it is interactive and seamlessly integrates virtual and physical elements. Overall, mixed reality scores 5 out of 5. Table 4.4 represents the comparison of advanced features of HoloLearn with other reality technologies.

Here is a comparison of the four different types of reality in terms of immersion, interactivity, portability, and social interaction:

Immersion: Virtual reality and mixed reality offer the highest levels of immersion, as they are fully immersive environments that can completely block out the real world. Augmented reality offers a less immersive experience, as it overlays digital content onto the real world rather than replacing it entirely. Reality offers the lowest level of immersion, as it is the real world itself.

Interactivity: All four types of reality offer high levels of interactivity, as you can interact with the environment in some way. In reality, you can interact with the physical world around you. Virtual reality, augmented reality, and mixed reality offer additional ways to interact with the environment, such as through the use of controllers or hand gestures.

Portability: Reality is the most portable of the four, as it can be experienced anywhere. Augmented reality is also portable, as it can be experienced through a smartphone or other device that you can take with you. Virtual reality and mixed reality are less portable, as they typically require specialized equipment that may not be as easy to take with you.

Social interaction: Reality offers the highest level of social interaction, as you can interact with other people in the physical world. Virtual reality, augmented reality, and mixed reality offer varying levels of social interaction, depending on the technology being used and the design of the experience.

Table 4.4 Comparison of advanced features of HoloLearn with other reality technologies.

Reality	Immersion	Interactivity	Portability	Social interaction
Reality	Low	High	High	High
Virtual reality	High	High	Low	Medium
Augmented reality	Medium	High	High	Medium
Mixed reality	High	High	Low	High

[8] Alahmari K, Duh H, Skarbez R. Outcomes of virtual reality technology in the management of generalised anxiety disorder: a systematic review and meta-analysis. Behaviour & Information Technology 2022:1–13.
[9] Farashi S, Bashirian S, Jenabi E, Razjouyan K. Effectiveness of virtual reality and computerized training programs for enhancing emotion recognition in people with autism spectrum disorder: a systematic review and meta-analysis. International Journal of Developmental Disabilities 2022:1–17.
[10] Merlo EM, Myles LAMK, Pappalardo SM. The VESPA project: virtual reality interventions for neurocognitive and developmental disorders. Journal of Mind and Medical Sciences 2022;9(1):16–27.
[11] Bekele E, Zheng Z, Swanson A, Crittendon J, Warren Z, Sarkar N. Understanding how adolescents with autism respond to facial expressions in virtual reality environments. IEEE Transactions on Visualization and Computer Graphics 2013;19(4):711–20.
[12] Cheng Y, Bololia L. The effects of augmented reality on social skills in children with an autism diagnosis: a preliminary systematic review. Journal of Autism and Developmental Disorders 2023:1–15.
[13] Hashim HU, Yunus MM, Norman H. 'AReal-Vocab': an augmented reality English vocabulary mobile application to cater to mild autism children in response towards sustainable education for children with disabilities. Sustainability 2022;14(8):4831.
[14] Root JR, Cox SK, Davis K, Gonzales S. Using augmented reality and modified schema-based instruction to teach problem solving to students with autism. Remedial and Special Education 2022;43(5):301–13.
[15] Mosher MA, Carreon AC, Craig SL, Ruhter LC. Immersive technology to teach social skills to students with autism spectrum disorder: a literature review. Review Journal of Autism and Developmental Disorders 2022;9(3):334–50.
[16] Kim K, Billinghurst M, Bruder G, Duh HB-L, Welch GF. Revisiting trends in augmented reality research: a review of the 2nd decade of ISMAR (2008–2017). IEEE Transactions on Visualization and Computer Graphics 2018;24(11):2947–62.
[17] Shahmoradi L, Rezayi S. Cognitive rehabilitation in people with autism spectrum disorder: a systematic review of emerging virtual reality-based approaches. Journal of NeuroEngineering and Rehabilitation 2022;19(1):91.
[18] Howlin P, Gordon RK, Pasco G, Wade A, Charman T. The effectiveness of Picture Exchange Communication System (PECS) training for teachers of children with autism: a pragmatic, group randomised controlled trial. Journal of Child Psychology and Psychiatry 2007;48(5):473–81.
[19] Kaale A, Smith L, Sponheim E. A randomized controlled trial of preschool-based joint attention intervention for children with autism. Journal of Child Psychology and Psychiatry 2012;53(1):97–105.
[20] Rossignol DA, Frye RE. Mitochondrial dysfunction in autism spectrum disorders: a systematic review and meta-analysis. Molecular Psychiatry 2012;17(3):290–314 [ASDSD].
[21] Philip RCM, Dauvermann MR, Whalley HC, Baynham K, Lawrie SM, Stanfield AC. A systematic review and meta-analysis of the fMRI investigation of autism spectrum disorders. Neuroscience & Biobehavioral Reviews 2012;36(2):901–42.

CHAPTER

Wearable sleeve for physiotherapy assessment using ESP32 and IMU sensor

5

M.N. Sumaiya, G.S. Vachanamruth, Vaishnavi Naveen, C. Varshitha and V.P. Yashaswini

Department of Electronics and Communication Engineering, Dayananda Sagar Academy of Technology and Management, Bangalore, Karnataka, India

1. Introduction

Physiotherapy is considered a science that uses physical activity to heal people. This autonomous medical profession is a nonsurgical procedure aimed at relieving pain, restoring function, and improving the ability of the patients to move the affected area of the body. A severe surgery is followed by prolonged rehabilitation depending on how the patient follows exercises on a regular basis [1]. Patients those who are suffering from joint fractures and dislocations must frequently see a physiotherapist and keep up a consistent exercise routine. This is a very time-consuming and expensive treatment, as the current digital devices used to test the range of motion (RoM) are both expensive and challenging to use. It is difficult to have constant access to physiotherapists, as they are very less in number. As per the Ministry of Health and Family Welfare statistics, India has only 0.59 physiotherapists for every 10,000 patients with a total population of 1.37 billion. In this regard, Norway topped having three professionals per 1000 patients. Moreover, physiotherapists are not available in rural areas. Waiting time at hospitals is lengthy and inadvisable for elder persons.

1.1 Existing methods

An instrument called a goniometer is used to gauge a joint's RoM. Goniometry is the art and science of determining the joint ranges in each plane of the joint [4]. Physical therapists typically use a goniometer to measure range of motion. The therapist can use a goniometer to measure RoM in a particular joint if a patient or client has decreased RoM in that joint. The goniometer can then be used in the following sessions to see whether the intervention is effective. The most used form of goniometer is the universal goniometer. Short-arm and long-arm versions of the universal goniometer are available. For joints with long levers, such as the knee and hip joints, the

Computational Intelligence and Deep Learning Methods for Neuro-rehabilitation Applications
https://doi.org/10.1016/B978-0-443-13772-3.00002-9

long-arm goniometer provides more accurate results. For smaller joints, such as the wrist, elbow, or ankle, the short arm goniometer is utilized effectively. The use of a digital goniometer has a number of advantages, including accessibility, measuring simplicity, measurement tracking via applications, and one-handed use to compute the joint angles. But the currently used goniometer is very expensive.

1.2 Scope and objectives

A technical solution to assist patients, is to remotely track the patient's motion as suggested by the physiotherapists. For that, a survey has been conducted by consulting physiotherapists and other professionals in the field to get a grasp of the actual requirements. From the insight obtained, the priority lies with the problem of ease of obtaining range of motion, for which a market research was conducted for the hardware components that can be used. Inertial measurement unit (IMU) sensor is used to track patient movement and transmit them as input to Node MCU. Node MCU is used for processing and transmission of data to the computing unit. Normalized human limb exercise data are collected to create a database for progress tracking. RoM data for patient from wearable sleeve are to be updated in the database for comparison. An algorithm for comparing data with previous analysis and monitoring recovery is to be developed. A personalized report with a graphical study has to be generated.

2. Literature survey

In Ref. [1], the paper describes the existing methods of calculation of the RoM along with its limitations. Because of the time constraints in occupational medicine, a visual estimate of RoM is the most desired and often-used method. This method can be used for any joint and does not require any additional equipment; however, it may produce misleading results due to inaccurate angular values while visualizing RoM. Universal goniometer (UG) is an equipment used to measure joint RoM. In both inter- and intraexaminer rehabilitation, it provides good accuracy and reliability. It is inexpensive, portable, and noninvasive. However, skilled professionals are required to evaluate this instrument. The paper analyses the relatively improved results when using the IMU sensor to track the motion of an individual, which gives the relative orientation of the part where the chip is placed. A wearable sleeve design is also incorporated along with visual reality to aid patients in performing the suggested exercises. However, in support of this proposed method are the quality metrics used, that is, the Kolmogorov—Smirnov (K—S test), a coherent method to check if two datasets vary from each other and the Bland—Altman plot. In this, two datasets have been considered healthy individual dataset and patient undergoing recovery dataset. In comparison to the current methods used, this system is cost efficient and simpler to use.

In Ref. [2], the paper discusses the limitations of performing sessions without a skilled professional. Patients may, for example, perform their exercises incorrectly if they are not supervised by their therapist. Wrong exercise alignment, wrong movement speed, and poor advancement quality all have the potential to diminish exercise efficacy and lead to a negative consequence. Patient compliance is another important issue with home-based rehabilitation exercises. Recent research has looked into how technology may be utilized to improve home-based exercise schedules with patients being encouraged to exercise at home by receiving supportive feedback. Patients have identified that biofeedback devices provide valuable information on exercise technique and precision, allowing them to adjust their motions in real time. It has been proposed that inertial sensors are used to track movement during exercise and provide information on exercise and accuracy. These sensors are great for measuring movement and providing feedback to patients while they conduct their exercises because of their small size and inconspicuous nature. Furthermore, because of the low cost and ease of use of these sensors, a number of inertial sensor-based systems have been developed to address the issues associated with home exercise therapy. Mainly, the paper researches the accuracy of using a single sensor to report the RoM. Recent research looked into whether decreasing the number of sensors used to decide exercise performance from three to two, then one on categorizing ability scores. Also, this study reveals that a single sensor can provide enough information on exercise performance in comparison with inertial sensors, thus by preventing the need for several sensor units.

In Ref. [3], the system methodology used in this paper is to interface a device with a Kinect sensor to monitor a patient's physiotherapy session, which provides feedback on the patient's performance through vibrohaptic pulses supplied from an armband, specifically looking for movements of the specified joint. An array of vibratory motors are in armband, which links to the KinoHaptics framework running on a computer via Bluetooth. In addition, the device uses a computer monitor to provide visual input. The KinoHaptics system provides feedback on the user's performance based on a workout program designed by a physiotherapist, and the KinoHaptics system loads this as a configuration file. During a physiotherapy session, KinoHaptics gives vibro-haptic comments whenever a predetermined angle of elevation for a certain joint (body part) is attained. In our proposed method, a database for both healthy individuals, as well as patients, is used instead of creating a configuration file.

In Ref. [4], the paper refers to elbow spasticity, a condition of muscle stiffness and use of wearable bands with motion detection and machine learning to rectify the condition. Spasticity is a common sign in patients suffering from neurological disorders. Upper and lower limbs' irregular movements are evaluated on a regular basis to determine functional outcomes of physical therapy, and the modified Ashworth scale-based clinical outcome measures are employed to quantify them. This research concentrated on analyzing acceleration and rotation aspects collected from the affected side of patients' elbows to determine the severity of elbow spasticity. Then the degree of spastic movement is classified using a machine-learning algorithm, which is similar to assigning an MAS score. Inertial data from individuals are collected during a passive stretch test by using a wearable device. Machine-learning methods were tested in combination with two segmentation methodologies and feature sets using decision trees, random forests (RFs), support vector machines, linear discriminant analysis, and multilayer perceptrons. Comparatively, RF performed well with 95.4% accuracy. It shows that wearable technology and machine learning can be used to create a clinically meaningful index, but also it enables rehabilitation patients to measure their spasticity levels in nonhealthcare situations. It is inferred that the inertial data can be processed for measuring the severity of spasticity.

In Ref. [5], the findings relevant to this project can be specified as the existing healthcare system is fundamentally expensive in terms of time, money, and resources. Directing research toward the development of novel e-Health solutions will aid in the provision of smarter and less expensive healthcare services. In fact, the existing approach to patients' care is ineffective. Therapy requires a rigorous schedule of daily, weekly, or bimonthly physiotherapy sessions in the case of patient rehabilitation in a hospital. Exercises must be done on a daily basis under the guidance of a doctor or other healthcare professional in the case of poststroke patients. The patient is being watched by the medical personnel in charge giving the patient appropriate feedback during exercise. The disadvantage of this method is that the patient must travel to the hospital for the training session which causes discomfort to persons who have motion limitations. In reality, physically disabled patients dread spending time in public areas because they are obliged to reveal their limitations to others. Exercising at home gives more psychological comfort to patients due to lower mental pressure and increased confidence. Patients can also create their own training timetable, allowing them to better organize their time.

Other flaws in the traditional healthcare system are related to expenses: each transfer to the hospital for a rehab session incurs additional expenditures including private or public transportation. Furthermore, this plan requires doctors and other healthcare professionals to be physically present during exercise, necessitating longer shifts and, as a result, higher compensation. The IoT-based vision allows patients to train individually, be evaluated, and receive real-time information from the remote system. Furthermore, because physicians and physical therapists are in short supply, the remote self-training technique can avoid lengthy waiting periods and waiting lists. By analyzing various studies, it is apparent that remote monitoring solutions are vital to the new eHealth architecture's efficacy. As a result, the extant literature provides a number of choices for doing real-time human monitoring, some of which are presently commercially available.

In Ref. [6], pose estimate is useful in a variety of industries, including film-making, sports biomechanics, and physical rehabilitation. The patient's initial assessment is mostly qualitative and subjective in rehabilitation at present. The therapists have very fundamental instruments at their disposal, such as goniometry, which is a technique for determining the size of a single human joint. When the patient is moving, the angle cannot be determined. The therapists must track the patient after the initial examination, make progress and double-check that patients have completed all of the exercises they have been given correctly. The latter is particularly challenging because multiple patients may need to be supervised at the same time, and this occurs frequently. Patients are given activities to complete on their own time. The aim of this survey is to create a system that uses small, nonintrusive inertial measuring units to properly determine the patient's lower body kinematics (positions, velocities, and accelerations of joint angles). Continuous position measurements allow extracting data such as range of motion, number of strides, and striding length during gait therapy. IMUs were the most preferred one than sensors, for physiotherapy applications due to their compact size and inexpensive cost. IMUs allow several patients to be in close proximity without any interruption.

The purpose of the research outlined in Ref. [7] is to develop a wearable, low-cost smart knee monitoring system. In this work, data from multiple sensors (IMU, temperature, pressure, and GSR—galvanic skin reaction) are fused to demonstrate the proposed wearable monitoring system. By monitoring rotation and acceleration, IMU sensors provide mechanical data like knee angle, motion, and direction. Other low-cost, low-power, compact, and light sensors report and monitor knee skin conductance, temperature, and muscle pressure around the knee joint. The pressure sensors are mounted on the inside of the movable armband. Two extra IMUs with two Velcro straps are attached on the inner side of an adjustable and ergonomic textile-based knee brace. GSR sensor measures local skin sweat gland activity (skin conductance) by using skin tension and perspiration. As the activity of the sweat glands increases, so does the conductance of the skin. Before extracting knee angles from accelerometer measurements, the IMU's orientation vector in a stationary position has to be determined on the shank and thigh with respect to the Earth frame.

In Ref. [8], the fore and upper arm movements are measured and evaluated by using two IMU sensors, one mounted inside the M5 Stick and the other linked to the extension port. The fore and upper arm angles are used to compute the elbow aperture angle. The I2C protocol is used to communicate with the external IMU. The designed electro goniometer has a tiny form factor and sends data to a computer through Bluetooth. As a result, it can be attached to the patient's limbs and used to obtain measurements while allowing them to move around freely. A graphical user interface (GUI) is included with the hardware, which may record measurements for subsequent study in real time for visual feedback. The patient can use the 3D simulation for assessment during therapy sessions at home. The pitch, roll, and yaw angles are calculated using quaternion coefficients to eliminate the gimbal lock phenomenon, which is a source of error for IMU sensors during angle measurements. To increase the accuracy and stability of the observations, the collected angle values are filtered using the Kalman filter. The correlation coefficient of system measurements is found to be quite high when compared to traditional goniometer data. In comparison to prior research, the created electro goniometer and its associated GUI improve the usability of the device while also providing high accuracy.

In Ref. [9], the objective of this work is to present all features by the unique Kinect-based Tele rehabilitation System. It handles two new capabilities that may be incorporated into them, in addition to the features supplied by existing software. The potential to retrieve information from patients' physical therapy records and treatment regimens is addressed. TRHONT is an ontology that helps doctors find the right exercises for their patients. The tele immersion feature makes the experience simple, effective, and user-friendly. A two-way real-time multimedia communication system is employed for offering tele-rehabilitation.

In Ref. [10], remote monitoring of physical training sessions could provide data on training outcomes to physicians and physical therapists, which could be useful in customizing workouts. Patients will be able to get better rehabilitation outcomes in less time as a result of this. IoT, virtual reality serious games, and wearable sensor network are employed to increase patient involvement and to track their progress during physical rehabilitation. Diverse virtual reality scenarios in games allow patients to do exercises with a highly interactive and nonintrusive approach using a set of wearable gadgets, thus aiding rehabilitation motivation. One of the body's most intricate joint systems is the shoulder complex, which includes the sternoclavicular, scapulothoracic, acromioclavicular, and glenohumeral joints. Triplanar upper limb movements are made possible by the shoulder's design. These movements produce kinematics that cannot be precisely captured when compared to conventional mechanical or robotic joints. For these reasons, quantifying upper limb kinematics is the most challenging task of estimating human motion.

The measurement of shoulder movements is crucial for evaluating upper limb movement, which is obviously clinically significant; hence this obstacle must be overcome. If affordable, accessible IMUs could take the role of pricy, inconvenient kinematic labs, physicians would be better able to comprehend the effects of their interventions in routine clinical practice [11,12]. Since IMUs have been used widely to estimate joint kinematics, IMU-based kinematic models are constructed using machine learning techniques. But, it estimates the orientation of the IMU by combining accelerometer and gyroscope data, estimating the orientation of the anatomical segments by using a sensor-to-segment calibration, and calculating joint angles as the relative orientation between reference frames fixed to adjacent body segments [13–15]. Wearable IMUs are used for measuring joint angles which overcome these drawbacks. Recent research [16–20] conclude that IMU-based kinematic models of lower-limb activities achieve absolute differences relative to optoelectronic-based models.

Considering these issues, the solution is proposed to track and monitor the progress and rehabilitation of patients who are in need of physiotherapy. The proposed work addresses a wearable digital gadget design that measures RoM. A user interface that encourages patients to exercise and tracks their development in a simple and practical way without requiring frequent hospital visits. Patients can view their progress and results graphically. The proposed cost-effective and wearable portable sleeve is consisting of ESP32 as the microcontroller and IMU sensor (accelerometer, gyroscope, and magnetometer) for measuring angular and planar limb motion of patients.

3. Proposed system: Materials and methodology

3.1 IMU sensor

IMU is a sensor that detects magnetic field, force, and occasionally angular rate. IMUs are referred to as 6-axis IMUs which are made up of 3-axis accelerometer and 3-axis gyroscope. Third 3-axis magnetometer is referred to as 9-axis IMU. IMUs can be utilized to measure acceleration, angular velocity, magnetic fields, and motion, orientation. IMUs are used with sensor fusion software, which combines data from several sensors to provide estimates of orientation and heading [3]. They are used in a wide variety of consumer electronics and industrial applications. These motion sensors can be used to calculate the angle measurements of the exercises. There are three types of motion sensors combined in a single IMU unit. The accelerometer is the motion sensor type that is most frequently employed. It is not advised as a distance measurement technique because of the double integration and the status of current technology. Gyroscopes use three axes of pitch (x axis), roll (y axis), and yaw to detect angular velocity. The orientation of an item in 3D space can be determined by a gyro when it is combined with sensor fusion software. Whenever gyroscope lacks an initial frame of reference, angular position can be calculated by combining its data with an accelerometer. A magnetometer measures magnetic fields by monitoring the magnetic flux density of the air at the sensor's location in space and can be identified with the variations of the Earth's magnetic field. Absolute heading can be calculated using this in combination with accelerometer and gyroscope data.

3.2 System architecture

System architecture outlines the system's structure, behavior, and additional viewpoints. An architecture description is specified to facilitate analysis of its components and behavior. System components used in the entire system are included in system architecture. System architecture block diagram is shown in Fig. 5.1.

The physiotherapy assessment system consists of two distinct but interconnected units, the Wearable Control Unit and the Database unit [5]. The wearable unit is present at the sleeve whereas the database is implemented on the server system.

1. Wearable control unit: The wearable control unit is an efficient, low-cost, low-power consumption, and smart accessible system used in the tracking and recording aspect of the physiotherapy assessment system. It consists of an ESP32 set up with the Espressif library using micropython Thonny IDE. The unit is connected with the MPU9250 sensor which is a 9 degrees of IMU. The readings from the MPU9250 nominally in raw format are processed through an algorithm to obtain the roll, pitch, and yaw for motion tracking. This processing is completed within the ESP32 similar in processing as that of the concept of edge computing before it is forwarded through the internet onto the database.

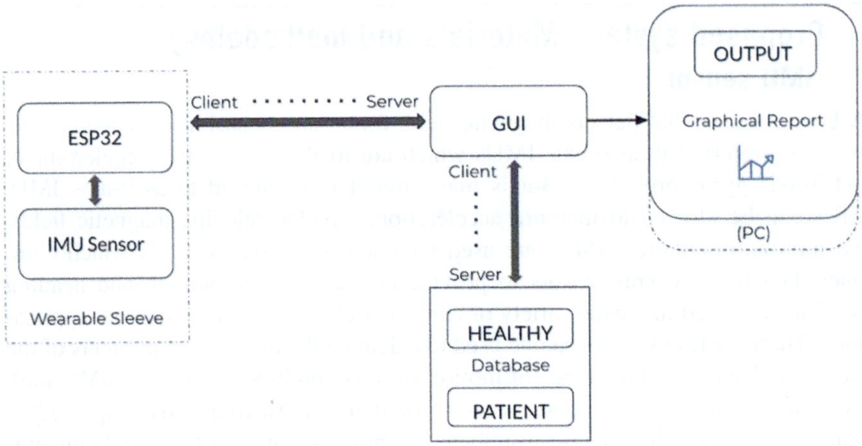

FIGURE 5.1

Block diagram of the proposed model.

2. Database unit: This is the software aspect of the system where the data obtained are stored. A client-server connection is set between the wearable unit and the database. The database stores two separate data for optimal comparison, one that of patient's and the other of healthy individuals. The stored data are further analyzed to present the graphical report.

3.3 System implementation

IMU sensor is used to calculate relative orientation in three-dimensional spaces, which contain an accelerometer, gyroscope, and magnetometer. Firstly, the IMU sensor is calibrated to prevent measurement uncertainty. Range of operations of IMU sensor is analyzed in all degrees of freedom, further a database is created, which stores the exercise details of healthy individuals. The readings of IMU sensor are obtained when it is worn by the patient while performing the exercise. Readings from IMU are sent to ESP32. Pitch and roll are measured using an accelerometer and gyroscope reading [6,7].

$$\text{PITCH} = (-\text{ATAN } 2(\text{ACC}[0], \text{SQRT}(\text{ACC}[2] * 2 + \text{ACC}[1] * 2)) * \text{RAD\_TO\_DEG}) \tag{5.1}$$

$$\text{ROLL} = (\text{ATAN } 2(\text{ACC}[1], \text{SQRT}(\text{ACC}[2] * 2 + \text{ACC}[0] * 2)) * \text{RAD\_TO\_DEG}) \tag{5.2}$$

The values or pitch and roll are calculated using the formula in Eqs. (5.1) and (5.2), respectively, where ACC[0], ACC[1], and ACC[2] are the values of acceleration due to gravity in x, y, and z directions. ATAN2() functions calculate the arctangent of x and y/x, respectively, where x and y are the coordinates of a point (x, y). The ATAN2() function returns a value in the range $-\pi$ to π radians. Yaw is calculated using a magnetometer. Pitch, roll, and yaw along with patient details are inserted into the patient's database using client–server connection. Further, a graphical report of the patient is generated after 10 days by comparing records of the patient in the patient's database with the record of the healthy individual database corresponding to the exercise that the patient has performed. This report helps the patient to understand their progress. Along with the graphical report, the percentage of recovery is also displayed. The system was implemented on a Node MCU ESP32 Board, which has in-built WiFi and Bluetooth module. ESP32 runs on breakout boards and modules from 160 to 240 MHz, and it has dual-core 32-bit microprocessors, which run up at very high speed to 600 Dhrystone Million Instruction per second.

3.4 Installation of mircopython on ESP32

MicroPython is a reimplementation of the Python 3 programming language designed for microcontrollers and embedded systems. Regular Python and MicroPython are extremely similar. With a few exceptions, MicroPython has all of the same language features as Python. The fact that MicroPython was created to function in limited circumstances is the main distinction between Python and MicroPython. In this system, we use Thonny IDE to program the ESP32 with Micropython programming language. The ESP32 or ESP8266 boards do not come preflashed with MicroPython. Flashing, uploading, or burning the firmware is the first step in MicroPython board programming. It can be accomplished in a variety of ways. Thonny IDE comes with a tool that allows for quick installation of MicroPython firmware on your board, then interpreter ESP32 is connected through COM port.

3.5 Interfacing MPU9250 with ESP32

MPU9250 is a 9-axis motion processing unit that has reduced chip size and power consumption, and it integrates the latest InvenSense design innovations. It leads to improvement in performance and low cost. The schematic circuit connection for the sleeve is shown in Fig. 5.2. The pin connection of MPU9250 and ESP32 is as mentioned in Table 5.1.

3.6 Graphical user interface

The GUI that the patient is able to view on their PC is both a client and a server. It takes the role of a client to the main server that handles the angle values coming from the GUI and enters the same into the tables in the database. The GUI acts as a server for the ESP32 as it sends the commands to start the recording of the angle values of the exercise and receive the maximum angle value of that set of a particular exercise.

Once the GUI is launched, it goes to a standby screen waiting for the ESP32 to establish its connection to the GUI, Fig. 5.3A. When the ESP32 is connected to the GUI, clicking the "Check for connection" button leads to the next screen, Fig. 5.3B. In the next screen, Fig. 5.3B, the patient is prompted to enter their details which are

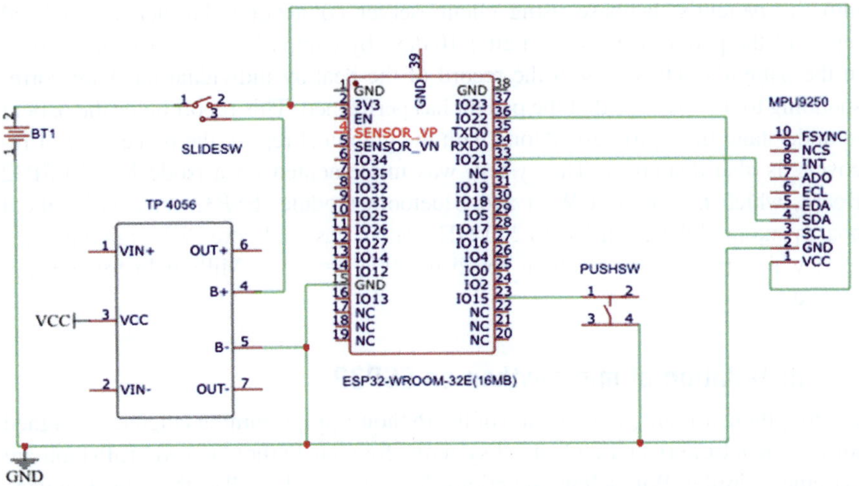

FIGURE 5.2

Circuit schematic for wearable sleeve.

Table 5.1 Connection between MPU9250 and ESP32.

ESP32 connection	MPU9250 connection
Vin: 3.3 V/5 V	3.3 V
GND	GND
Send and receive data (SDA-21)	Send and receive data
Clock signal (SCL-22)	Clock signal

the following: Name, Phone Number, and Age. Along with these details, the patient is required to select the exercise prescribed to them by their doctor and the number of sets of that exercise. Upon filling the above details and clicking the "Submit" button, the next screen is loaded, Fig. 5.3C. The screen shown in Fig. 5.3C informs the patient the exercise they have chosen to perform and the status of the sets yet to be performed, that is, either pending or completed. To start a set of the chosen exercise, clicking the "Start set" button informs the ESP32 to go on standby mode waiting for the patient to press the button on the sleeve to start the exercise at their convenience. Once the button on the sleeve is pressed, the ESP32 collects the MPU9250 sensor's reading. Upon completion of the set, the ESP32 informs the GUI the same and transmits the angle value of that set along with the completion signal. The same procedure is repeated until all the sets have been performed. After the successful completion of all the sets, the data can be uploaded to the server by clicking the "submit" button.

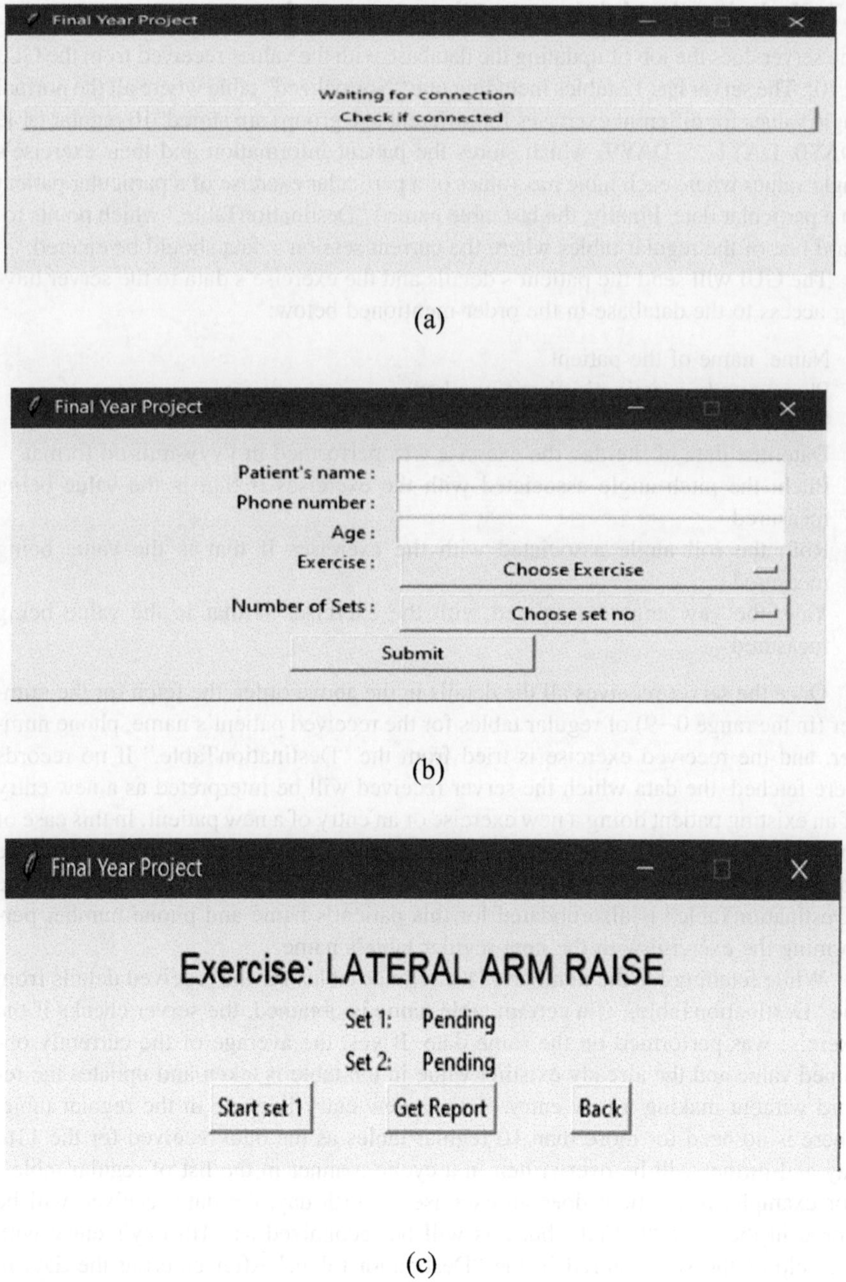

FIGURE 5.3

Graphical User Interface (A) waiting for connection (B) enter patient details (C) perform exercise.

3.7 Updating the database on the server end

The server does the job of updating the database with the values received from the GUI [9,10]. The server has 12 tables including one "Normalized" table where all the normal angle values for different exercises for different age groups are stored. 10 regular table (DAY0, DAY1 … DAY9) which stores the patient information and their exercise's angle values where each table has values of a particular exercise of a particular patient on a particular date. Finally, the last table named "DestinationTable," which points toward one of the regular tables where the current session's data should be entered.

The GUI will send the patient's details and the exercise's data to the server having access to the database in the order mentioned below:

- Name: name of the patient
- Phone number: patient's phone number
- Exercise: the exercise the patient is performing
- Date: the data of the day the exercise was performed in yyyy-mm-dd format
- Pitch: the pitch angle associated with the exercises if that is the value being measured
- Roll: the roll angle associated with the exercises if that is the value being measured
- Yaw: the yaw angle associated with the exercises if that is the value being measured

Once the server receives all the details in the above order, the fetch for the number (in the range 0–9) of regular tables for the received patient's name, phone number, and the received exercise is tried from the "DestinationTable." If no records were fetched, the data which the server received will be interpreted as a new entry of an existing patient doing a new exercise or an entry of a new patient. In this case of a new entry, the server updated the "DAY0" (one of the 10 regular tables) table in the database with all the details in the same order mentioned above. Along with this, the "DestinationTable" is also updated for this patient's name and phone number performing the exercise with the next regular table's name.

While fetching for the number of the regular tables for the received details from the "DestinationTable," if a certain table name is obtained, the server checks if the exercise was performed on the same date. If yes, the average of the currently obtained value and the already existing value in the table is taken and updates the record without making a new entry, if no, a new entry is made in the regular table. There is no need for more than 10 regular tables as the data received for the 11th day and further will be overwritten in a cyclic manner in the list of regular tables. For example, if a patient does an exercise on 11th day, the data received will be stored in the table "DAY0," but this will be recognized as 11th day's entry with the help of the value stored in the "DestinationTable." After entering the data in the "DAY0" table, the server updates the next destination, that is, "DAY1" for 12th day; for this, patient performing this exercise as 11. During the next entry for the same patient's details on a different day, the server performs a modulus 10 ($11\%10 = 1$) operation on the value received and gives the value as 1, using which the server updates the "DAY1" table.

3.8 Detecting the angle of the set

As soon as the ESP32 receives a signal from the GUI to start a set, it enters into a standby mode. At this stage the patient is required to wear the sleeve on the limb at an appropriate position and press the button provided on the sleeve to start the set. Once the button is pressed, the patient is required to start their exercise. The ESP32 now records the MPU9250 sensor values for 40 s at an interval of 0.5 s.

After the completion of the recording, the maximum value of the required planar angle is identified. Now all the values recorded are compared to check if they fall in between the range of maximum value and 20 degrees below the maximum value. If the values satisfy the said condition, they are recorded into a new list, for example, list A. As soon as the condition is not met, the recorded list is appended to another list, for example, list B, and list A is emptied. This procedure is repeated till all the recorded values are checked against the mentioned condition. Finally, list B takes the form of a list of lists.

The length of list B gives us the number of repetitions of the exercise in that particular set. Each list in list B is browsed through to get the value of the maximum angle in that particular repetition. The average of the maximum angles of all the repetitions is taken and sent back to the GUI as the angle value of the set along with the set completion signal.

3.9 Graphical report

If a patient or their doctor intends to view their progress of a particular exercise in a graphical format, it can be obtained by clicking the "Get Report" button in Fig. 5.3C. To get the report, the patient must have performed the exercise for at least 10 days. This minimum limit is set to get meaningful insight from the report. The report consists of a graph that plots the angles on the *y*-axis performed by the patient for a particular exercise against the days on the *x*-axis. Along with the patient's angles, in the same graph, there is also a plot of angles by a normal person, in the same age group, who is able to perform the same exercise. This graphical report is intended to show visuals of the patient's progress over the past 10 days and how much more is required to reach the normal angle performed by a normal person. Along with the graph, the percentage of the progress over 10 days and the percentage of recovery remaining are shown.

4. Experimental results and discussion

The prototype of the proposed wearable sleeve shown in Fig. 5.4 was implemented with the assistance of two patients who had ailments in their shoulders. These patients were prescribed two different exercises, namely Frontal arm raise and Lateral arm raise.

FIGURE 5.4

Wearable sleeve prototype.

The front raise is shown in Fig. 5.5A; this isolation exercise is recommended for patients who are recovering from a shoulder injury or shoulder surgery. It strengthens the shoulder muscles and the upper chest and reduces neck pain. The lateral arm raise, shown in Fig. 5.5A, is useful in strengthening shoulder movement and effective in isolating the lateral head muscle with regular slots.

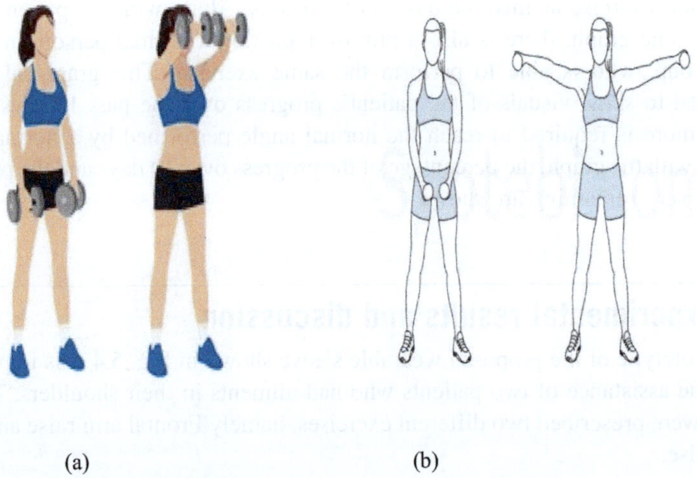

FIGURE 5.5

(A) Frontal arm raise; (B) lateral arm raise.

4.1 Quantitative measures

The subjects performed their prescribed exercises wearing the system/sleeve for 10 consecutive days, as shown in Fig. 5.6, and the graphical report of their progress is as shown in Fig. 5.7.

4.2 Quality measures

The data used to plot the graph shown in Fig. 5.7 are given in Tables 5.2 and 5.3. The roll and yaw in the experimental data are a constant zero, respectively, as the exercise prescribed to those patients required only one plane's angular value. Table 5.2 shows the data of subject 1, aged 70, who was prescribed "Frontal arm raise," which was performed on the given dates. Table 5.3 shows the data of subject 2, aged 69,

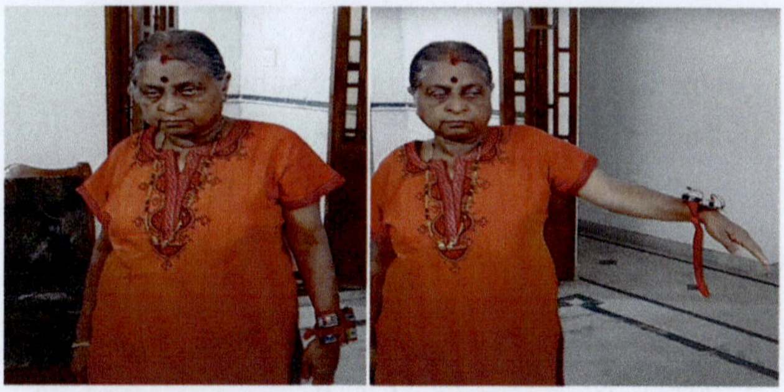

FIGURE 5.6

Subject 1 performing exercise.

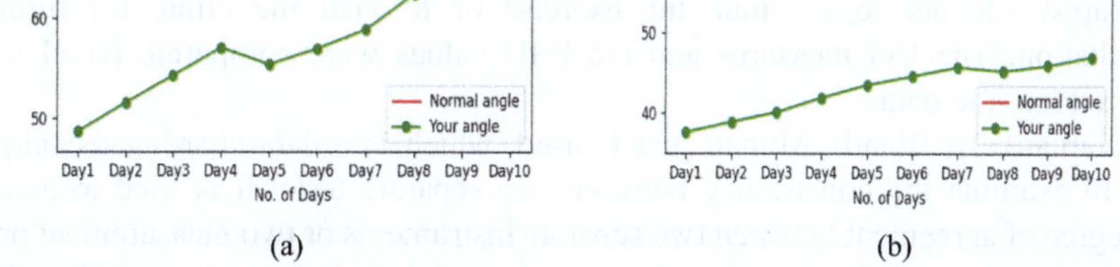

FIGURE 5.7

Graphical report shown to the patient/doctor: (A) front arm raise; (B) lateral arm raise.

Table 5.2 Experimental data of subject 1.

Date	Pitch
2022-5-18	48.7
2022-5-19	51.6
2022-5-20	54.3
2022-5-21	57.1
2022-5-22	55.4
2022-5-23	57.0
2022-5-24	58.9
2022-5-24	62.5
2022-5-26	64.8
2022-5-27	67.7

Table 5.3 Experimental data of subject 2.

Date	Pitch
2022-5-18	37.6
2022-5-19	38.9
2022-5-20	40.1
2022-5-21	41.9
2022-5-22	43.5
2022-5-23	44.6
2022-5-24	45.7
2022-5-24	45.2
2022-5-26	45.9
2022-5-27	47.0

who was prescribed “Lateral arm raise,” which was performed on the given dates. It can be seen that there is a gradual improvement in the range of motion over 10 days.

Progress reports of exercise frontal/lateral arm exercise can be seen by both patients and physiotherapist all the time. Suggestions can be given to patients by physiotherapist whether to continue the exercise or to visit the clinic for further consultation. The UG measures and the IMU values were compared. Excel was used to store the data.

Alternatively, Bland–Altman plot is used, which is a data charting technique used to examine the consistency between two separate tests. It is used to assess the degree of agreement between two separate instruments or two measurement procedures. Many researchers have used this plot to validate their results with the other method. But, a major disadvantage is that a strong correlation does not always indicate that the two approaches are in good agreement with one another. The central bias line displays the standard deviation between measurements and whether both measurement methods can be used interchangeably with a 95% probability of producing values that are similar because the variations between them are in the upper and lower limit of agreement. The difference between these and the central bias suggests that both systematic and random errors are present. These mistakes are not a result of measurement technologies; rather, human error and the constraints of the measurement process are to blame. Any deviation from the upper and lower limit of agreement (less than a 95% probability of comparable readings) for that measurement indicates a lack of device agreement. But in our proposed wearable device, a graphical report will be generated, and the chart will be compared with their previous exercise movements.

5. Conclusion and future scope

The project intends to provide a portable, cost-efficient, and rapid physiotherapy assessment to patients. It provides precise monitoring report of patient's activities during the session. This wearable sleeve is aimed to assist old and sick people in rural areas to understand the correctness of the exercises performed by them without the presence of a physiotherapist. The development of this project is an effective tool for patients to understand the correctness of the exercises performed by them. As using multiple sensors has been in the norm for motion tracking, this project uses a single sensor to evaluate performance. Our project incorporates a user-friendly GUI easily accessible by the patients keeping their comfort in mind. The data gathered are quantified and published as a report along with the graph symbolizing their development to the patient helping them monitor their conditions effectively.

Recently, a demand has been raised for motion-tracking devices in which use of wearable motion capture systems to measure ROM. Portable, low-cost, and user-friendly IMUs are used in widely adapted motion tracking devices. In this chapter, the wearable sleeve is proposed to track and record the physiotherapy assessment system. It is used to assess the treatment smartly in different angles using ESP32, MPU 9250 sensor, and IMU sensor. Acceleration, angular velocity, and magnetic inclinations are measured by IMU–MPU9250 sensor to track human motion. Connection between the wearable unit and the database is set and analyzed to display the graphical report. The GUI acts as a server for the ESP32 to send and record the angle values of the exercise.

This proposed wearable sleeve can be worn on the limb at an appropriate position and press the start button. Once the exercise is started, ESP32 records the MPU9250 sensor values for 40 s at an interval of 0.5 s. The graphical report shows the progress of patient's over the past 10 days and further the progress of the patient is calculated and plotted in the form of graph.

Acknowledgments

The authors would like to thank the anonymous editors and reviewers for their productive comments to raise the quality of the chapter. Also, this work is funded and supported by VTU Sponsored Student Project, Karnataka, India and part of project under VGST-KFIST L1, Karnataka, India with the grant number GRD 786.

References

[1] Narejo A, Baqai A, Sikandar N, Ali A, Narejo S. Physiotherapy: design and implementation of a wearable sleeve using IMU sensor and VR to measure elbow range of motion. International Journal of Advanced Computer Science and Applications 2020;11:445. https://doi.org/10.14569/IJACSA.2020.0110953.

[2] Giggins OM, Sweeney KT, Caulfield B. Rehabilitation exercise assessment using inertial sensors: a cross-sectional analytical study. Journal of NeuroEngineering and Rehabilitation 2014;11:158. https://doi.org/10.1186/1743-0003-11-158.

[3] Rajanna V, Vo P, Barth J, Mjelde M, Grey T, Oduola C, et al. KinoHaptics: an automated, wearable, haptic assisted, physiotherapeutic system for post-surgery rehabilitation and self-care. Journal of Medical Systems 2016;40:1−12.

[4] Kim J-Y, Park G, Lee S-A, Nam Y. Analysis of machine learning-based assessment for elbow spasticity using inertial sensors. Sensors 2020;20:1622. https://doi.org/10.3390/s20061622.

[5] Bisio I, et al. Towards IoT-based eHealth services: a smart prototype system for home rehabilitation. 2019 IEEE Global Communications Conference (GLOBECOM). 2019. p. 1−6. https://doi.org/10.1109/GLOBECOM38437.2019.9013194.

[6] Joukov V, Karg M, Kulic D. Online tracking of the lower body joint angles using IMUs for gait rehabilitation. 2014 36th Annual International Conference of the IEEE Engineering in Medicine and Biology Society. 2014. p. 2310−3. https://doi.org/10.1109/EMBC.2014.6944082.

[7] Faisal AI, Majumder S, Scott R, Mondal T, Cowan D, Deen MJ. A simple, low-cost multi-sensor-based smart wearable knee monitoring system. IEEE Sensors Journal March 15, 2021;21(6):8253−66. https://doi.org/10.1109/JSEN.2020.3044784.

[8] Keskinoğlu C, Aydın A. Wearable wireless low-cost electrogoniometer design with Kalman filter for joint range of motion measurement and 3D modeling of joint movements. Proceedings - Institution of Mechanical Engineers H February 2021;235(2):222−31. https://doi.org/10.1177/0954411920971398. Epub November 13, 2020. PMID: 33183138.

[9] Anton D, Berges I, Bermúdez J, Goñi A, Illarramendi A. A telerehabilitation system for the selection, evaluation and remote management of therapies. Sensors 2018;18:1459. https://doi.org/10.3390/s18051459.

[10] Postolache O, Hemanth DJ, Alexandre R, Gupta D, Geman O, Khanna A. Remote monitoring of physical rehabilitation of stroke patients using IoT and virtual reality. IEEE Journal on Selected Areas in Communications Feb. 2021;39(2):562−73. https://doi.org/10.1109/JSAC.2020.3020600.

[11] Rigoni M, Gill S, Babazadeh S, Elsewaisy O, Gillies H, Nguyen N, et al. Assessment of shoulder range of motion using a wireless inertial motion capture device-a validation

study. Sensors April 13, 2019;19(8):1781. https://doi.org/10.3390/s19081781. PMID: 31013931; PMCID: PMC6514956.

[12] Höglund G, Grip H, Öhberg F. The importance of inertial measurement unit placement in assessing upper limb motion. Medical Engineering & Physics 2021;92:1−9. ISSN 1350-4533.

[13] Bailey CA, Uchida TK, Nantel J, Graham RB. Validity and sensitivity of an inertial measurement unit-driven biomechanical model of motor variability for gait. Sensors 2021;21:7690.

[14] Wei W, Kurita K, Kuang J, Gao A. Real-time limb motion tracking with a single IMU sensor for physical therapy exercises. 2021 43rd Annual International Conference of the IEEE Engineering in Medicine & Biology Society (EMBC). 2021. p. 7152−7. https://doi.org/10.1109/EMBC46164.2021.9630480.

[15] de Villa SG, Martín AJ, Domínguez JJG. Adaptive IMU-based calibration of the center of joints for movement analysis: one case study. In: 2020 IEEE International Symposium on medical measurements and applications (MeMeA); 2020. p. 1−6. https://doi.org/10.1109/MeMeA49120.2020.9137135.

[16] Komaris D-S, Tarfali G, O'Flynn B, Tedesco S. Unsupervised IMU-based evaluation of at-home exercise programmes: a feasibility study. BMC Sports Science, Medicine and Rehabilitation 2022;14(1):1−12.

[17] García-de-Villa S, Jiménez-Martín A, García-Domínguez JJ. Novel IMU-based adaptive estimator of the centre of rotation of joints for movement analysis. IEEE Transactions on Instrumentation and Measurement 2021;70:1−11.

[18] Yi C, Rho S, Wei B, Yang C, Ding Z, Chen Z, Jiang F. Detecting and correcting IMU movements during joint angle estimation. IEEE Transactions on Instrumentation and Measurement 2022;71:1−14.

[19] Doğan M, Koçak M, Kılınç ÖO, Ayvat F, Gülşah S, Ayvat E, et al. Functional range of motion in the upper extremity and trunk joints: nine functional everyday tasks with inertial sensors. Gait & Posture 2019;70:141−7. https://doi.org/10.1016/j.gaitpost.2019.02.024. ISSN 0966-6362.

[20] Zhou L, Fischer E, Tunca C, Brahms CM, Ersoy C, Granacher U, Arnrich B. How We found our IMU: guidelines to IMU selection and a comparison of seven IMUs for pervasive healthcare applications. Sensors 2020;20(15):4090. https://doi.org/10.3390/s20154090.

CHAPTER

6

Machine learning for developing neurorehabilitation-aided assistive devices

Shivalika Goyal and Amit Laddi

Imaging, Avionics and Display Systems Department, CSIR-CSIO, Chandigarh, Punjab, India

1. Introduction

The nervous system facilitates communication between all body components by using electrical and chemical mechanisms. It is the electrical wiring of the body. The brain, spinal cord, and a network of nerves having special cellular structures called neurons that connect to every region of the body make up this structure. The nervous system is in charge of receiving and processing sensory data, managing movement and coordination, and regulating a number of bodily processes like digestion, blood pressure, and heart rate [1]. This system is essential for healthy operations of the body and any disruptions or damage to it can have serious repercussions. Numerous factors, such as trauma, illness, and hereditary abnormalities, can lead to nervous system dysfunction. According to an analysis of World Health Organization (WHO), the burden of fatalities and disabilities brought on by neurological illnesses is becoming more widely acknowledged as a worldwide public health concern, and over the next several decades, as the population ages, this burden is expected to increase [2].

One of the most popular frameworks providing information on summary assessments of population health across illness and risk categories is the global burden of disease (GBD) method. The GBD framework uses a single metric to characterize the burden of diseases from diagnostic categories of the International Classification of Diseases (ICD) [2]. According to the GBD framework, neurological illnesses account for 12% of all worldwide fatalities and 6.3% of the global disease burden. Neurological illnesses or disorders are the second greatest cause of mortality and the primary cause of disability globally. When calculated as a fraction of the "total years of healthy life estimated to have been lost" due to impairments resulting from neurological illnesses, as well as other diseases and traumas with neurological sequelae, accounted for 14% of the total in 2005, and this percentage is expected to be the same by 2030 [3].

Computational Intelligence and Deep Learning Methods for Neuro-rehabilitation Applications
https://doi.org/10.1016/B978-0-443-13772-3.00007-8

1.1 Neurological disorders

Neurological disorders are conditions that affect the functioning of the brain, spinal cord, and/or the nerves that connect them, and it can be fatal if proper procedures and medicines are not provided on time [4,5]. The absolute number of disability-adjusted life years (a measure of the burden of neurological illnesses globally) is rising [6]. Governments will have to deal with a rise in demand for treatment, rehabilitation, and support services for neurological illnesses as populations are rising and aging and significant debilitating neurological conditions are more common as people become older. There are many different types of neurological disorders, ranging from relatively mild conditions to more serious and debilitating disorders. Some common categories of these disorders from which neuro disabilities arise include:

- **a)** *Neurodevelopmental disorders:* These disorders are characterized by delays or abnormalities in the development of the nervous system and can include conditions such as autism and intellectual disability [4].
- **b)** *Neuropsychiatric disorders:* These disorders affect the mind and behavior and can include conditions such as anxiety disorders, mood disorders, and psychotic disorders [7].
- **c)** *Neuromuscular disorders:* These disorders affect the muscles and the nerves that control them and can include conditions such as muscular dystrophy and myasthenia gravis [8].
- **d)** *Neurovascular disorders:* These disorders affect the blood vessels that supply the brain and can include conditions such as stroke and aneurysms. A stroke is a medical emergency that causes the destruction of brain cells when the blood supply to the brain is disrupted [9].
- **e)** *Neurodegenerative disorders:* These disorders involve the progressive loss of function in the nervous system and can include conditions such as Parkinson's disease and Alzheimer's disease. They both are progressive brain disorders in which the earlier impacts are on the movement and later to the memory, cognition, and behavior [9].
- **f)** *Traumatic brain injury:* This type of neuro disorder can occur as a result of a traumatic event, such as a car accident or a fall where the injury is brought on by a blow to the head or a wound that penetrates the skull. It can cause a range of symptoms depending on the severity of the injury [10].

The above-mentioned examples of neurological disorders might have several additional conditions that can impact the nervous system. Multiple factors including genetics, infections, trauma, and underlying medical diseases like diabetes or high blood pressure can contribute to the development of neurological disorders. Depending on the illness, there are a variety of ways to treat neurological problems including drugs, counseling, surgery, assistive technology, and lifestyle modifications.

1.2 Neurorehabilitation and assistive technology

Neurorehabilitation is a type of therapy that helps individuals with neurological disorders or injuries to regain or improve their cognitive, physical, and/or sensory functioning. It is a multidisciplinary division where a variety of experts such as physicians, physical therapists, occupational therapists, speech therapists, and psychologists collaborate to assist people obtain the best results. The goals of neurorehabilitation vary depending on the specific needs of the individual, but may include improving mobility, communication skills, cognitive function, and independence in daily activities [11]. Along the continuum of care from the acute hospital to the community, rehabilitation entails a coordinated and iterative problem-solving process. Physical medicine, pharmacology and nutrition, psychology and behavior, education and counseling, occupational and vocational guidance, social and supporting services, architecture and engineering, and other treatments are examples of specific rehabilitation approaches. Rehabilitation services act as a link between exclusion and isolation. Rehabilitation is a potent instrument to give self-empowerment [12]. Neurorehabilitation can be provided in a variety of settings including inpatient rehabilitation facilities, outpatient clinics, and in the home. It can involve a range of interventions, such as the following.

a) *Physical therapy:* to improve mobility, balance, and coordination

b) *Occupational therapy:* to help individuals perform daily activities and return to work

c) *Speech therapy:* this type of therapy focuses on improving communication skills, including speech, language, and swallowing.

d) *Cognitive therapy:* this type of therapy focuses on improving cognitive functions such as memory, attention, and problem-solving skills.

e) *Behavioral therapy:* this type of therapy focuses on addressing and changing any emotional or behavioral issues related to the neurological condition that may be problematic or unhealthy, such as substance abuse or aggression.

f) *Recreational therapy:* this type of therapy uses leisure activities and hobbies to help individuals with disabilities improve physical, mental, and social skills.

g) *Assistive technology:* this type of therapy involves the use of devices and technology to help individuals with disabilities perform tasks that may be difficult or impossible for them to do on their own. These may include devices such as wheelchairs, braces, and splints to assist with mobility or communication devices to help with speaking and understanding language [11,13].

The specific type of neurorehabilitation that is most appropriate for an individual depends on their specific needs and goals. Various neurological disorders such as multiple sclerosis, Parkinson's disease, brain damage, spinal cord injury, stroke, and others can benefit from neurorehabilitation. Although it can be useful at any point in the recovery process [14]. It is frequently most beneficial when it is initiated as early as possible after the injury or the commencement of the condition. In case of any neurological disorder detection after receiving primary care, patients are counseled to enroll in a long-term rehabilitation program. The specific treatment plan is tailored to the individual's needs and may involve a combination of different approaches. This generally includes regular hospital visits and an activity schedule done at home. According to studies, frequent changes and the associated difficulty of commuting to and from the rehabilitation facilities cause many patients to withdraw from the outpatient rehabilitation program. The distribution of rehabilitation facilities is likewise very uneven. Large cities are typically home to a lot of such centers whereas there are less, or no such establishments exist in rural areas. These are a few of the main challenges that patients seeking to take part in a rehabilitation program need to be worked upon. The need of the era is for technology to enable routine rehabilitation treatment to be delivered in patients' comfortable homes with minimal clinical intervention derived from assistive technology in neurorehabilitation [13].

Assistive technology can be referred to a technology or apparatus used to assist such impaired individuals in carrying out their daily duties and activities. It can take many different forms and may be valuable for those with physical, sensory, or cognitive impairments. People with these impairments can benefit greatly from assistive technology since it can enable them to execute necessary tasks more easily and independently. Tethering these individuals to social, professional, and educational opportunities can also be positive lessoning for them to enhance their quality of life. There are several conventional assistive technologies that are commonly used in neurorehabilitation to help such individuals with neurological disorders or injuries improve function and independence. Some of them are wheelchairs, braces and splints, communication devices like speech-generating devices, hearing aids, adaptive equipment (adapted utensils, adapted clothing), prosthetics (artificial limbs), orthotics (devices that support or correct the function of a limb or joint), lift chairs, adaptive beds, bikes, toys, augmentative and alternative communication (AAC) devices, and environmental control systems that include devices that allow individuals to control their environment, such as lights and appliances using a computer or other device [15]. This is significant but not limited as there are numerous other assistive devices that can be used in neurorehabilitation. The use of specific devices depends on the requirements and objectives of the individual. In the field of assistive technology for neurorehabilitation, machine learning techniques can be employed to create and enhance tools and technologies that allow people with neurological impairments or injuries to regain function and improve their standard of living. In this chapter, we elaborate on intelligent assistive technologies integrated with machine learning that have been effectively used in neurorehabilitation [16,17].

2. Machine learning as a tool in assistive technology for neurorehabilitation

Machine learning (ML) is a subset of artificial intelligence (AI) that involves the use of algorithms and statistical models to enable computers to learn and improve their

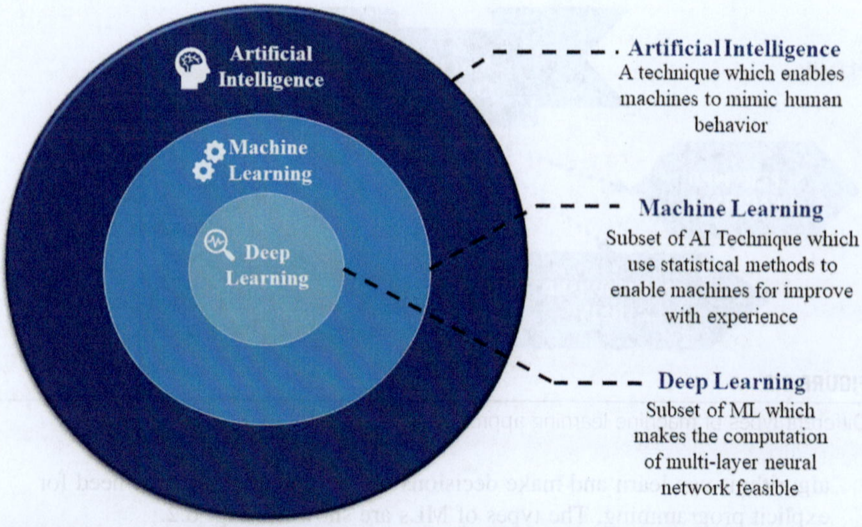

FIGURE 6.1

A quick lookout of AI, machine learning, and deep learning.

performance over time without being explicitly programmed. It has the potential to be a useful tool in neurorehabilitation, particularly in the use of assistive technology and development of personalized devices that can adapt to an individual's changing needs and abilities [18]. A brief lookout of AI and ML is shown in Fig. 6.1.

ML is a method of data analysis that automates analytical model building. It is a branch of AI that focuses on the design and development of algorithms that can learn from and make predictions on data. In ML, an algorithm is trained on a data set. The algorithm then makes predictions or decisions without being specifically programmed to perform the task. For instance, an ML model may be used to evaluate mobility, which might improve at-home therapy while simultaneously lowering recurring costs. However, such systems need intensive training and testing [19,20]. Any ML system must be trained on reliable, unbiased data.

There are several different types of ML approaches, including:

a) *Supervised learning:* In this type of ML, the algorithm is trained on labeled data meaning that the data set includes both input data and the corresponding correct output. The algorithm makes predictions based on this input–output mapping.

b) *Unsupervised learning:* In this type of ML, the algorithm is not given any labeled training data. Instead, it must find patterns and relationships in the data on its own.

c) *Reinforcement learning:* In this type of ML, an agent learns through trial and error receiving rewards or penalties for certain actions.

d) *Deep learning:* This type of ML involves the use of artificial neural networks which are inspired by the structure and function of the brain. Deep learning

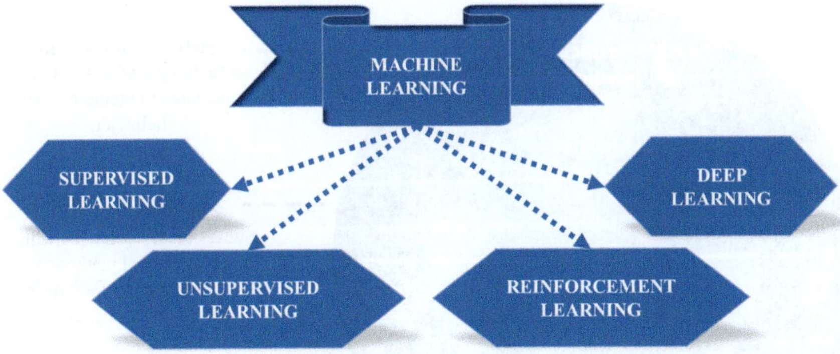

FIGURE 6.2

Different types of machine learning approaches.

algorithms can learn and make decisions on their own without the need for explicit programming. The types of MLs are shown in Fig. 6.2.

Machine learning has a wide range of applications, including image and speech recognition, natural language processing, and predictive modeling. It is being used in a variety of industries, including healthcare, and has the potential to revolutionize neurorehabilitation [21].

In the field of assistive technology for neurorehabilitation, ML techniques can be used to develop and improve devices and technologies that help individuals with neurological disorders or injuries to regain function and improve their quality of life. One potential application of ML in neurorehabilitation is in the development of personalized assistive devices that can adapt to an individual's changing needs and abilities [22]. For example, ML algorithms could be used to analyze data from sensors and other sources to understand an individual's patterns of movement and activity, and then use this information to adjust the settings or features of an assistive device in real time to better meet the individual's needs. This is an intelligent prosthetics and exoskeleton that can adapt to the user's movements and provide feedback and guidance to help individuals learn new skills and improve their mobility. Intelligent assistive technologies adapt to the user's changing needs and abilities over time improving their effectiveness and user experience [23]. ML algorithms are used to analyze large datasets of brain activity, movement, and other data to better understand how the brain and body functions and how they can be aided through assistive technology is another application area. There are several other ways in which ML can be used in assistive technology including speech recognition allowing individuals who are unable to type or write to communicate more easily, image recognition helping individuals with visual impairments allowing them to recognize and interpret their surroundings, predictive text input to facilitate them to predict words that they are likely to type making it easier for them to communicate, natural language processors where ML algorithms support to understand and interpret natural language allowing individuals to communicate with assistive technology using their own voice, and much more [22].

Overall, the use of ML in assistive technology for neurorehabilitation has the potential to greatly enhance the effectiveness and usability of these technologies and to improve the lives of individuals with neurological disorders or injuries. It has the potential to improve the effectiveness and efficiency of assistive technologies and rehabilitation programs and to help individuals with neurological disorders or injuries to achieve their maximum potential for recovery and independence [24,25].

2.1 Integration of ML in assistive technology and its applications

The process of using ML techniques in integration with any assistive technology typically follows a common workflow that involves the following steps as shown in Fig. 6.3 and later discussed in detail:

a) *Defining the problem/task:* The first step in any ML plan is to define the problem that one is trying to address. This involves identifying the goals like the learning to be integrated to the assistive device embedded with an ML algorithm, the data availability, and the specific ML techniques to be used.

b) *Collecting and preparing data:* The next step is to collect and prepare the data that will be used to train the ML model. This may involve cleaning and preprocessing the data, as well as selecting relevant features and splitting the data into training and testing sets [26].

c) *Choosing and configuring a ML model:* Once the data is prepared, one needs to choose an ML model that is appropriate for the task at hand. This involves selecting a model architecture and configuring the model with appropriate hyperparameters.

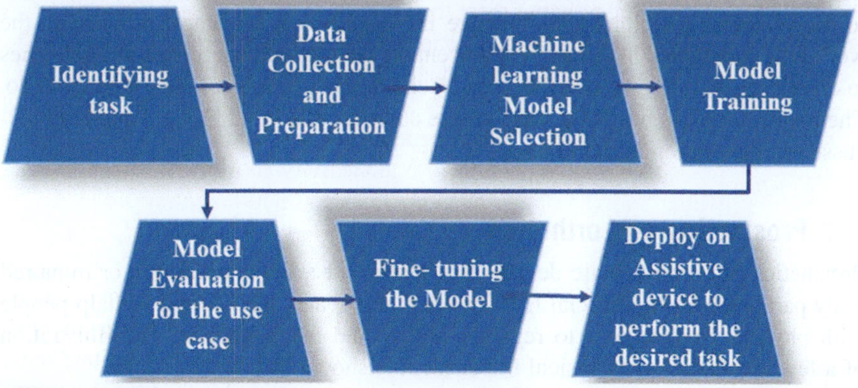

FIGURE 6.3

Common workflow for the integration of ML with assistive technology.

d) *Training the model:* Once the model is configured, one can begin training it on the prepared data. This involves using an optimization algorithm to minimize the loss function of the model which measures how well the model is able to make predictions on the training data.
e) *Evaluation of the model:* After the model is trained, one needs to evaluate its performance on unseen data to determine how well it is able to make accurate predictions. This may involve calculating various performance metrics, such as accuracy, precision, and recall.
f) *Fine-tuning the model:* If the model's performance is not satisfactory, one needs to go back and fine-tune the model by adjusting the hyperparameters, adding or removing features, or using different model architectures.
g) *Deploying the model:* Once the model is performing well, one can deploy it in a production environment which is the assistive device to begin using it to make predictions or take other desired actions. This trained model makes the device intelligent.

This workflow can be iterative with the model being trained and evaluated multiple times until it is performing well enough for the task at hand [27].

3. Applications of ML-integrated assistive technology in neurorehabilitation

ML algorithms are used to create tools and gadgets that help in neurorehabilitation, and this approach is described as ML-integrated assistive technology. Such technologies aim to restore or enhance the cognitive, physical, and behavioral abilities of individuals with neurological disorders or impairments. ML in neurorehabilitation assistive technology has the potential to increase the efficacy and efficiency of rehabilitation services and to provide individualized and accessible rehabilitation for patients with a variety of neurological illnesses and traumas. There are numerous forms of assistive technologies that are integrated with ML capabilities with the work folio mentioned in Section 2.1 and enable people with neurological disabilities to carry out tasks that would otherwise be challenging or impossible for them to do. The major instances based on aids and use cases are shown in Fig. 6.4 and later discussed in the section in detail.

3.1 Prosthetics and orthotics

Prosthetics and orthotics are devices that replace or support a missing or impaired body part, such as an artificial limb or a leg brace, and can be used to help people with physical impairments to regain mobility and independence. The illustration of a few prosthetics for physical impairment is shown in Fig. 6.5.

Even though orthotics and prosthetics are distinct from one another, they work together to promote recovery. They are considered a type of assistive technology

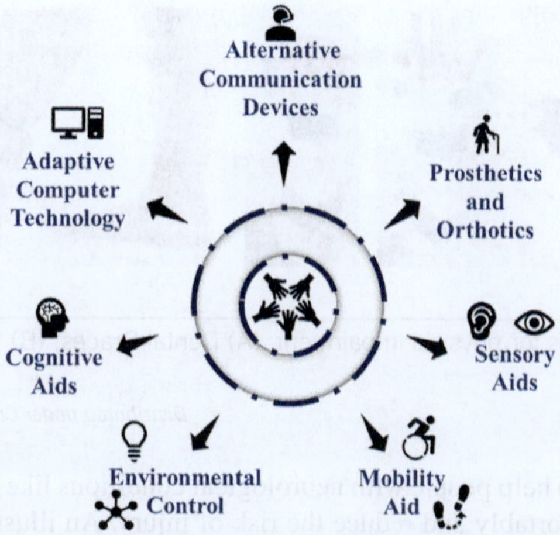

FIGURE 6.4

Various instances of ML based on aids and use cases.

FIGURE 6.5

Instances of prosthetics for physical impairment.

because they are artificial devices that are used to help people with disabilities to live more independently and participate fully in society. They can restore mobility, sensory function, and other abilities that have been lost due to injury, illness, or congenital conditions. Prosthetics can be used to replace or enhance body parts that have been lost or damaged due to a neurological event. For example, a person who has lost the use of their arms due to a stroke may use a prosthetic arm to perform tasks like dressing or cooking [28]. A person who has lost the use of their legs due to a spinal cord injury may use a prosthetic leg to walk again.

Orthotics are specific devices that are used to support, align, or correct the function of the bones, joints, and muscles of the body. The different types of orthotics that can be used in neurorehabilitation include braces that can be worn around a joint or body part to provide support and stability. They can be used to help people with neurological conditions like cerebral palsy or multiple sclerosis to move more easily and comfortably. Splints are worn on the arms or legs to support and stabilize joints. They can be used to help people with neurological injuries or conditions like carpal tunnel syndrome to recover and regain function. Orthopedic shoes are special shoes that have been designed to provide support and alignment for the feet and ankles.

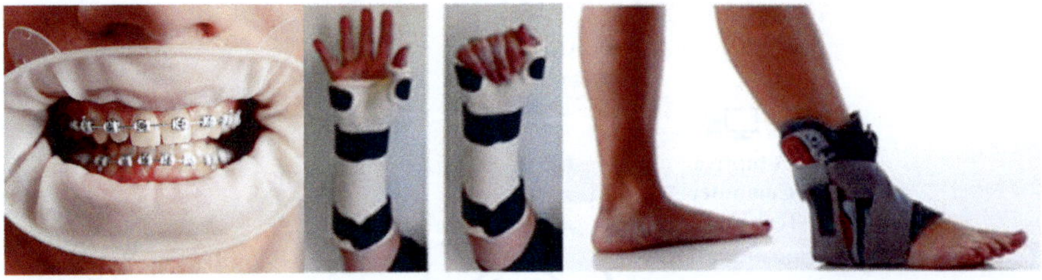

FIGURE 6.6

Illustration of orthotics for physical impairment. (A) Dental Braces, (B) Wrist Braces and (C) Splints.

They can be used to help people with neurological conditions like flat feet or bunions to walk more comfortably and reduce the risk of injury. An illustration of orthotics for physical impairment is shown in Fig. 6.6.

ML can be used in the design and development of prosthetics and orthotics to improve their performance and usability. There are several different types of ML models that can be used in the design and development of prosthetics and orthotics, including neural networks, support vector machines (SVMs), decision trees, and Gaussian process models [29,30]. These machine learning models have been used in prosthetics and orthotics to classify different types of movements, predict the intended movements of the user, for analysis based on past movements, and analyze sensor data respectively. Some ways in which ML is used include the following components:

a) *Movement prediction:* Movement prediction in the context of prosthetics and orthopedics involves developing algorithms and techniques to enable prosthetic devices to anticipate and respond to the movements of the user [31]. This can be achieved through a variety of approaches including ML techniques, sensor-based approaches, and electromyography (EMG) [32]. One approach to movement prediction in prosthetics involves the use of ML algorithms to analyze patterns of EMG activity and predict the intended movement of the user. This can be done by training a machine learning model on a dataset of EMG signals collected from the user while they perform various movements. The model can then be used to predict the intended movement of the user based on their current EMG activity [33]. Sensor-based approaches to movement prediction in prosthetics involve the use of sensors to detect the movements of the user and anticipate their intended actions. This can be achieved using sensors such as accelerometers, gyroscopes, and magnetometers which can be used to detect changes in the position and orientation of the prosthetic device. A technique for extracting important characteristics from multivariate signals such as electroencephalography (EEG) or EMG data is called the common spatial pattern (CSP) algorithm. Finding patterns in the brain or muscle activity that are connected to certain motor activities or cognitive processes is a prevalent practice in the area of neurorehabilitation. To distinguish the desired signal from background noise and other forms of interference, the CSP algorithm seeks out a collection of spatial filters. These filters are made to minimize the difference within each class while maximizing the difference between signals in two classes, such as right-hand movement and left-hand movement. To categorize the signals and extract pertinent characteristics for future analysis, the resultant pattern known as CSP components can then be utilized. The CSP algorithm's capacity to handle high-dimensional data, such as that collected from EEG recordings with several electrodes is one of its key features. It has also been demonstrated to be beneficial in numerous neuro-rehabilitation applications, including stroke rehabilitation and motor control, and is reasonably easy to put into practice [34]. Overall, the goal of movement prediction research in prosthetics is to enable prosthetic devices to respond more intuitively and naturally to the movements of the user [35]. Companies like ReWalk Robotics and Bioness work on similar lines.

b) *User customization*: ML algorithms can be used to develop personalized interventions for individual patients. Assistive technology can learn and adapt to the preferences and needs of the user over time. For example, the device could learn the user's preferred speed and acceleration, movement patterns, medical history, and other relevant factors and adjust its performance accordingly. This can be achieved using supervised learning in which the ML model is trained on a dataset of labeled information. The model can then be used to recommend specific interventions or exercises to help the patient [36].

c) *Obstacle avoidance*: Obstacle detection and avoidance in prosthetics involve the use of sensors and other technologies to identify and avoid obstacles that may be encountered by the user. This is an important consideration for people using assistive devices as they may have limited mobility and may rely on their devices to navigate their environment [37]. There are several approaches to obstacle detection in prosthetics, including the use of sensors such as LIDAR, ultrasound, and stereo cameras [38]. These sensors can be used to detect the presence and distance of objects in the environment allowing the prosthetic device to avoid colliding with them. ML algorithms are used to analyze sensor data and identify patterns that indicate the presence of an obstacle. Convolutional neural networks (CNNs) and recurrent neural networks (RNNs) are examples of deep learning algorithms that may be used to analyze sensory input from cameras, lasers, and other sensors and generate predictions about the environment. To train these algorithms to identify obstacles and provide suitable control signals, massive datasets of annotated pictures or point clouds can be used. The model can then be used to predict the presence of obstacles in real time based on the current sensor data [39].

d) *Fall prevention:* Fall prevention in neurorehabilitation involves developing techniques and technologies to help people with neurological conditions reduce their risk of falling and injuries related to falls. This can be a significant concern for people with conditions such as stroke, Parkinson's disease, and multiple sclerosis, as these conditions can affect balance, mobility, and coordination [40]. ML algorithms can be used to analyze sensor data and identify patterns that may indicate a fall is imminent, allowing the device to take preventative action. This can be achieved using supervised learning in which the ML model is trained on a dataset of labeled examples of movements with and without falls. The model can then be used to predict the likelihood of a fall based on the current movement data [41]. Another ML-based approach is by classifying motor task and marking the ones indicative of falling, where Spiking Neural Networks (SNNs can be used). Artificial neural networks that mimic brain cells' communications are known as SNNs. Unlike in a CNN network, an SNN uses pulses or "spikes" to communicate between neurons as opposed to continuous values. Due to its capacity to closely resemble some of the essential elements of biological neural networks, SNNs have been suggested as a viable method for neurorehabilitation. SNNs, for instance, may handle data in a more event-driven way which can be more effective and more suited to some jobs. In a real-world context where biological signals could be compromised, they can also be more resilient to noise and other types of interference. SNN research is currently underway, thus it is not yet known how well they will perform in neurorehabilitation applications as compared to more conventional approaches. To ascertain the possible advantages and disadvantages of SNNs for neurorehabilitation and to create efficient algorithms and methods for doing so, more study is required [42].

3.2 Augmentative and alternative communication devices

Augmentative and alternative communication (AAC) devices can be used as part of neurorehabilitation to help people with neurological conditions or injuries to communicate more effectively and participate more fully in social interactions and activities. AAC devices can be particularly useful for people who have difficulty speaking or communicating due to a neurological condition or injury [43]. For example, a person who has a stroke and difficulty speaking may use a speech-generating device to communicate, or a person with cerebral palsy may use a communication board with pictures or symbols to express their needs and wants.

There are many different types of AAC devices including picture and symbol-based communication boards as shown in Fig. 6.7 or cards that show pictures or symbols representing different words or ideas. A person can point to a picture or symbol to communicate their message. Speech-generating electronic devices that can produce synthesized speech or recorded messages [44]. A person can use a speech-generating device to communicate by pressing buttons or using a switch to select prerecorded or synthesized words or phrases. Text-to-speech software can be installed on a computer or mobile device and used to type out messages that are then spoken aloud by the device. Sign language-based assistive framework of hand gestures and body movements used to communicate. Some people who have difficulty speaking may use sign language as an alternative way to communicate [20]. The majority of these employ a backend machine learning strategy.

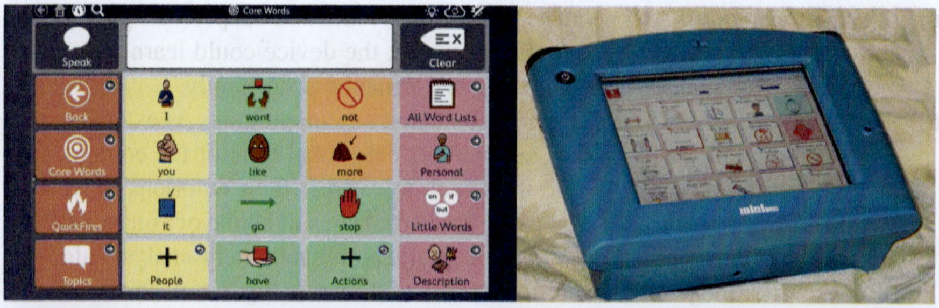

FIGURE 6.7

Alternative communication devices sample symbol-based communication boards.

In the context of AAC devices and neurorehabilitation, ML algorithms can be used to predict which words or phrases a person is likely to want to use based on their previous communication patterns so that the AAC device can present those options more quickly and accurately [43,44]. It can be of service in learning a person's unique speech patterns and vocal characteristics so that a speech-generating device can more accurately mimic their natural speech. Later it can also be significant in analyzing data on a person's progress over time to identify trends and patterns that may be useful in tracking their progress and adjusting their treatment plan [45].

There are various ML models used in the design and development of communication devices, such as AAC systems to improve their performance and usability. The algorithms of ML that can be used include but are not limited to:

a) *Natural language processing (NLP)*: NLP focuses on the use of natural language in interactions between computers and people. It entails creating program and systems that can decipher, interpret, and produce human language [46]. It enables users to speak directly to the device using their own voice. To create speech-generating technology that is more precise and responsive in translating a person's ideas or intents into spoken words, NLP might be employed as an example. NLP could be used to create tools that make it easier for persons with aphasia (a communication problem that can develop after a stroke) to comprehend and communicate [46,47].

b) *Text prediction:* Text prediction is an ML technique that anticipates the following word or series of words in a body of text. People with dyslexia or other reading disabilities may benefit from text prediction to make it easier for them to read and comprehend text [48]. For text prediction, several ML techniques can be utilized such as Markov models, *n*-gram models, and neural networks. These algorithms can be used to anticipate the following word or series of words in a particular context after being trained on massive textual datasets to understand linguistic patterns and structures [49].

c) *User customization:* Algorithms can be used to learn and adapt to the preferences and needs of the user over time. For example, the device could learn the user's preferred speech tempo and adjust its performance accordingly.

d) *Sentiment analysis:* The ML technique named sentiment analysis examines text or speech data to ascertain the sentiment or emotional tone of the content. Individuals with aphasia may find it difficult to express their emotions in words, but they may be able to do so by using nonverbal cues like tone of voice or facial expressions. These indicators could be analyzed by sentiment analysis algorithms to infer the person's emotional state as shown in Fig. 6.8. Decision trees, support vector machines, and neural networks are just a few of the machine learning methods that can be utilized for sentiment analysis [50]. These algorithms are trained on big volumes of tagged speech or text data to discover the patterns and characteristics linked to various emotions. The algorithms can be used to categorize new text or speech data as having a positive, negative, or neutral sentiment after they have been trained [51].

3.3 Adaptive computer technology

Software or hardware known as adaptive computer assistive technology is created specifically to make it easier for individuals with neurological conditions like dyslexia, ADHD, and autism to operate computers and other technology. This may involve tools like text-to-speech software, which reads text aloud for those who have trouble reading, or on-screen keyboards for those who have trouble typing [52]. Other examples of adaptive computer assistive technology include applications that help people with ADHD manage their time and stay organized as well as software that assists persons with dyslexia in editing their work. An example of this is shown in Fig. 6.9. Adaptive computer assistive technologies come in a wide variety and the ones that will be most beneficial will depend on the needs and capabilities of the user [53].

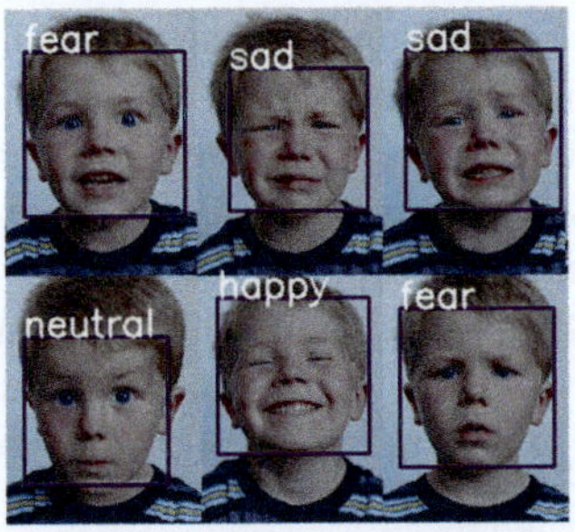

FIGURE 6.8

Sentiment analysis using machine learning approaches.

FIGURE 6.9

Adaptive computer assistive technology helping children with ADHD focus.

ML can be used in the design and development of computer access devices to improve their performance and usability. ML has a wide range of possible uses in adaptive computer assistive technology. Some illustrations include a device that employs machine learning to recognize and forecast when a person with ADHD is likely to become distracted and then sends notifications to encourage them to maintain focus [54].

A software that can analyze a user's social media activity using ML and offers feedback on any potentially harmful or inappropriate online conduct. ML is working as a tool to recognize and decipher the facial expressions and body language of people with autism giving them feedback to improve their comprehension of and decipher other people's emotions [50]. Predictive text input, image recognition, adaptive speech recognition, and user customization are a few examples of machine learning techniques that can be applied.

a) *Models for adaptive speech recognition systems:* Hidden Markov models (HMMs) which are based on the notion that the observable data, in this case the spoken words, are generated by a series of hidden states are just one of the many types of ML models that can be employed for speech recognition among others. HMMs are frequently employed in voice recognition because of their effectiveness and simplicity. Neural networks, which are modeled after the structure and operation of the human brain, are capable of discovering patterns and connections in data on their own. Due to their propensity for processing vast amounts of data and their capacity for adapting to new input, neural networks are frequently utilized for speech recognition [55,56]. Gaussian mixture models (GMM) presume that the data were produced by combining several distinct distributions, each of which is represented by a Gaussian curve [57]. To simulate the distribution of acoustic properties in speech signals, GMMs are frequently used in speech recognition [58]. SVMs are algorithms that locate the hyperplane in a high-dimensional space that best separates various kinds of data. SVMs have been applied to the classification of various phonemes or speech sounds, in speech recognition [59,60].

b) *Models for predictive text input:* For predictive text input, DL models, a form of artificial neural network can be utilized. RNNs, which are designed to analyze sequential input such as text and can learn to predict the next word in a sequence based on the preceding words, are some examples of DL models that could be used for this task [61,62]. RNNs of the sort known as long short-term memory (LSTM) networks were created with the objective of handling long-term dependencies in data. By learning to retain and use context from earlier words in the sequence to produce more precise predictions, LSTM networks have been used for predictive text input [63,64]. Transformer models are a sort of neural network that processes sequential data by applying self-attentional mechanisms. Transformer models are a sort of neural network that processes sequential data by applying a self-attentional mechanism [65]. By training themselves to pay attention to pertinent words in the input sequence and using this knowledge to create predictions, transformer models have been exploited for predictive text input [66].

c) *Models for image recognition systems:* There are several different types of ML models that can be used for image recognition, including convolutional neural networks (CNNs), a type of neural network that is specifically designed to process data, which has a grid-like structure such as an image [67]. Given their ability to scale to big data sets and their capacity to learn features directly from the data, CNNs are frequently utilized for image recognition [68]. By describing images as high-dimensional feature vectors and categorizing them based on the values of these features, SVMs have been used to recognize images. Instances are defined by *k*-nearest neighbor (KNN) models depending on how the majority of its nearest neighbors in the feature space are classed [69]. By describing images as feature vectors and determining the closest neighbors in the feature space to create a classification, KNN has been used to recognize photos [70]. Information can be classified using decision trees, a form of the tree-based model, depending on the values of different attributes. By classifying an image based on its pixel values and other properties, decision trees have been leveraged for image recognition [71].

3.4 Mobility aids

Mobility aids are a kind of assistive technology that can be especially beneficial for people with neurodevelopmental problems like autism, ADHD, and dyslexia as well

FIGURE 6.10
Smart wheelchair and cane as mobility aids for neurorehabilitation.

as for those who have had a stroke or brain injury [72]. These conditions may impair a person's ability to move around and navigate their surroundings which may make it challenging for them to engage in routine activities or reach locations [73]. These difficulties can be lessened, and a person's independence and movement enhanced with the aid of mobility aids [74]. The illustration of this is shown in Fig. 6.10.

Some examples of mobility aid that people with neurological problems may find useful are wheelchairs that make it easier for people to move around, particularly if they have trouble standing or walking. Wheelchairs come in a wide variety of designs, including manual wheelchairs, which the user propels themselves, and electric wheelchairs that are driven by an electric motor. Walkers and other gadgets offer assistance and stability to people who have trouble walking or keeping their balance. Walkers can be very beneficial for people who have cerebellar ataxia or Parkinson's disease [75,76]. Canes, which do not require the same level of support as walkers but can help those who have trouble walking balance and support.

Canes can be very useful for the people who have slight walking imbalance or one-leg weakness. While scooters are larger and more stable than electric wheelchairs, they also are termed power-operated vehicles (POVs). People who have trouble walking or standing still for extended periods of time frequently use them. In general, mobility aids can be a useful resource for people with neurodevelopmental disorders or other mobility disabilities enabling them to move around more easily and take part in daily life to a greater extent. The effectiveness and efficiency of mobility assistance devices for neurorehabilitation can be increased using ML algorithms. For example, to provide feedback and direction during rehabilitation activities, ML models, like CSP, SNN, and SVM, might be used to interpret data from wearable sensors such as motion sensors or EMG sensors [77]. This might make it easier for the therapist to focus on particular muscle groups or body parts that might benefit from extra care or training [78]. Some specific ways that ML models could be applied in the context of neurorehabilitation with mobility assistance devices include identifying patterns of muscle activity or movement that are indicative of development or improvement and providing feedback to the user or therapist, examining data from wearable sensors to pinpoint body parts or particular muscle groups that may require more training or attention, supplying individualized training regimens or rehabilitation plans based on each person's particular requirements, and monitoring a person's development over time modifying the treatment plan as necessary [77,79].

There are numerous different types of ML models that can be implemented for this kind of task learning, which include reinforcement learning models like Q-learning, deep Q-learning, policy gradient method, and actor-critic method, which involve interaction with the environment and receive rewards or penalties for particular actions, as well as learning to perform tasks through trial and error. A traditional RL approach called Q-learning includes learning a value function that calculates the predicted reward for doing a certain action in a specific condition. By altering the value function in response to observed rewards and state transitions, the agent learns to choose behaviors that maximize the predicted reward [80]. Neural networks are used for task learning by training the model on a large dataset of examples and enabling it to understand the underlying relationships in the data, whereas decision trees are used by examining the attributes of a task and deciding which actions to perform based on these aspects. Alternatively, K-means clustering is applied for task learning, which involves collecting related tasks into a single group and letting the model discover the underlying patterns in the data [81].

3.5 Sensory aids

Sensory aids are assistive devices that can help individuals with sensory impairments such as hearing loss or vision loss to more effectively perceive and interact with their environment. It can support neurorehabilitation efforts by boosting sensory input to the brain or making up for sensory impairments. Through the application of ML, sensory aids can become more efficient by being able to adapt to and learn from each patient's individual needs and preferences [82]. An illustration of hearing and vision sensory aid is shown in Fig. 6.11.

The following are some examples of sensory aids with potential ML integration possibilities that might be applied in neurorehabilitation:

a) *Hearing aids*: Depending on the user's particular hearing loss and listening environment, ML algorithms like CNN, SVM, and decision trees may be used to modify the settings of a hearing aid. For example, the hearing aid may be taught to automatically modify the loudness or frequency range in accordance with the user's preferences and the degree of background noise [83]. Here, to increase the device's signal-to-noise ratio, noise reduction algorithms can be used to evaluate sensor data, identify background noise, and reduce it. An SVM might be taught to categorize various noises as "speech" or "nonspeech" in the setting of hearing aids to enhance the signal's understandability or a decision

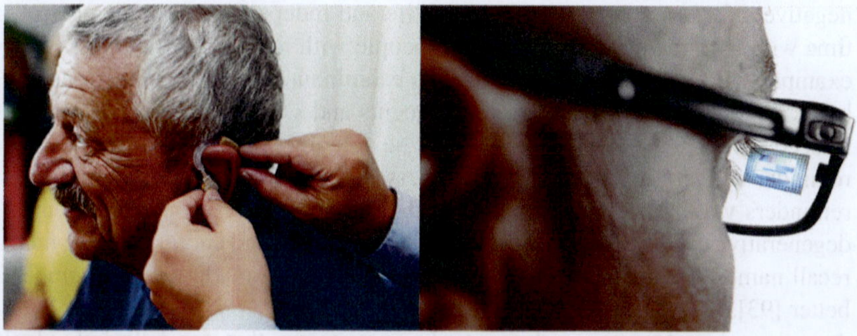

FIGURE 6.11

Sensory aid devices integrated with machine learning.

tree could be used to make a series of decisions about how to process an audio signal based on its characteristics, such as its frequency or intensity [84].

b) *Vision aids:* Text-to-speech software and low-vision spectacles might both benefit from improved picture recognition skills thanks to ML algorithms [85]. For example, based on the user's preferences and the lighting, the visual aid may learn to automatically modify the contrast or magnification. Here, ML techniques based on image enhancement may be used to evaluate and improve the clarity of pictures, enhancing the user's vision [86].

c) *Touchscreen devices:* ML techniques might be used to increase the usability of touchscreens for people with muscular disabilities [87]. For example, the system may be taught to detect the user's distinctive touch patterns and modify the sensitivity as necessary.

d) *User customization:* Algorithms for ML can be used to gradually learn and adjust to a user's choices and requirements. One possibility is for the gadget to figure out the user's preferred volume level and adjust appropriately.

3.6 Cognitive aids

Cognitive aids are tools or technologies that can assist those undergoing neurorehabilitation by enhancing cognitive abilities including memory, focus, and problem-solving abilities [88,89]. The following are some kinds of cognitive aids that might be applied in neurorehabilitation:

a) *Memory aids:* These can assist people with neurorehabilitation recall appointments, timetables, and other crucial information. These conventionally include memory books, calendars, and other tools. ML-based algorithms or models like CNN, SVM, and K-NN can be integrated with assistive devices to create memory aids (MS) for those with neurological diseases that impair memory such as Alzheimer's disease, Parkinson's disease, and multiple sclerosis [90,91]. These illnesses can lead to memory problems that have a substantial negative effect on a person's quality of life and independence. In the present time with ML, memory aids can assist people with memory difficulties. For example, ML algorithms may be used to examine a person's prior actions and behavior to forecast their future requirements and send warnings or reminders in accordance [92]. Additionally, it may be able to recognize a user's chosen reminder delivery mode (such as text message, email, or phone call) and send reminders via that means. It might possibly be helpful for people with neurodegenerative conditions by alerting them to key events and chores, helping them recall names and faces, or suggesting activities that can help them remember better [93].

b) *Attention aids:* Attention aids are tools or strategies that can help individuals with attention deficits improve their cognitive function [93–95]. These might be devices like white noise generators, fidget spinners, or stress balls to assist neurorehabilitation patients concentrate and stay on target. In the present time, ML-based smart watches, eye gaze trackers, and other wearable technology that may offer notifications and reminders can help neurorehabilitation patients stay focused and successfully manage their time. ML algorithms used in attention-assistive devices have the potential to assist people with neurodegenerative diseases by identifying patterns in their habits and offering recommendations to increase focus and attention [96,97].
c) *Problem-solving aids:* Games, puzzles, software programs, and other activities that might assist people undergoing neurorehabilitation refining over time and enhance their problem-solving abilities are referred to as problem-solving aids [96,98]. Tools and techniques for assisting people with neurological disorders that hinder problem-solving abilities, such as Alzheimer's disease, Parkinson's disease, and multiple sclerosis, can be developed in integration with ML models [99].
d) *Cognitive rehabilitation therapy:* A form of treatment called cognitive rehabilitation therapy, also referred to as cognitive remediation involves structured exercises and activities that are intended to enhance cognitive function [100]. A computer or other device can be used to offer customized cognitive rehabilitation therapy programs developed using ML algorithms. For example, a cognitive rehabilitation treatment software based on ML might be able to assess a person's cognitive capacities and create a personalized therapy plan that is suited to their requirements and objectives [101]. Additionally, the computer could be able to change the therapy plan in accordance with a person's development [102].

It is important to note that the use of ML in the development of memory, attention, and problem-solving aids is still a relatively new field of study where more research is required to assess how well these tools work for people with neurodegenerative illnesses. However, there is a big chance that ML might help people with neurodegenerative diseases live better lives. Thus, it is conceivable that more and more ML-based memory aids will be created in the future [98]. As it can assist people with neurological problems and improve their cognitive function and independence, cognitive aids and cognitive rehabilitation treatment are a significant component of neurorehabilitation [99]. In general, the use of ML in the construction of neurorehabilitation assistive technology has the potential to significantly raise the standard of living and increase the degree of independence of people with cognitive impairments.

3.7 Environmental control

With the use of environmental control assistive technology, people undergoing neurorehabilitation may better regulate their surroundings and can enhance their independence and quality of life [103]. To increase the functionality and usability of assistive environmental controls, machine learning may be applied in the design and development of these devices [104]. The following are some examples of environmental control assistive technology:

a) *Home automation systems:* People may manage their home's lighting, appliances, and other features with the use of a remote control, smartphone app, or voice commands. These systems may learn from and adjust to a person's habits and requirements owing to machine learning, enabling them to manage their home's lighting, utilities, and other features using a remote control, smartphone app, or voice commands [105]. It is also possible to employ machine learning algorithms to enhance the performance and general effectiveness of home automation systems. The system may, for instance, learn to reduce energy consumption by shutting off lights and appliances when they are not in use, or it could learn to identify and diagnose faults with systems and appliances to stop issues before they arise [106].

b) *Motorized equipment:* Various devices, such as motorized curtains or blinds, may be controlled by a remote control or other device and made intelligent to respond to the user's needs using machine learning [105,107]. This enables persons with neurorehabilitation to adjust their environment without expending physical effort.

c) *Adaptive switches:* With the development of sensor-based technology and their integration with machine learning methodologies, people with neurorehabilitation are now able to operate devices and appliances using a variety of techniques such as touch, breath, and eye gazing from switches that can be mounted on a wall or a tabletop [108,109].

Environmental control assistive technology can be an important part of neurorehabilitation, as it can help individuals with neurological impairments to live more independently and have more control over their environment.

4. Conclusion

This chapter comprised various aspects of the integration of machine learning with different types of assistive technologies used in aiding neurorehabilitation practices. In this chapter, we have concisely categorized the use of ML approaches in conventional, ongoing, and forthcoming tools of assistance exercised in physiological and behavioral therapies during rehabilitation for neurologically impaired individuals. Despite having significant and frequent technological advancements, some technologies need to have more suited implementations on the practical ground in real-time scenarios. Eye trackers and eyeglasses used as attentional aids in neurorehabilitation have limitations in terms of producing the same results as in a controlled lab environment due to the lack of sufficient, high-quality, nonrepetitive training data needed for the majority of optimized model design for machine learning approaches. Various machine learning approaches with advancements in hardware capabilities open prominent future possibilities for improvement and development of intelligent assistive technologies used in neurorehabilitation therapies such as gesture-controlled applications, embedded tools for mobility assistance, and edge computing for cognitive assistance tools. Researchers are posing their fullest contribution in designing and structuring ML-based technologies for neurorehabilitation along with its improvements in intelligence capacities and efficiency so that it can be used more resourcefully for the progressions on the industrial as well as public viable use cases. The goal of this chapter was to summarize different ML-based techniques for single sign-on understanding that can be advantageous for academic and research purposes.

References

[1] Cantile C, Youssef S. Nervous system. In: Jubb, Kennedy & Palmer's Pathology of domestic animals, vol. 1. Elsevier; 2016. p. 250–406.

[2] Mathers CD, Ezzati M, Lopez AD. Measuring the burden of neglected tropical diseases: the global burden of disease framework. PLoS Neglected Tropical Diseases 2007;1(2):e114.

[3] Sanjekar PS, Patil JB. An overview of multimodal biometrics. Signal Image Process 2013;4(1):57–64.

[4] Chaudhuri A, Behan PO. Fatigue in neurological disorders. The Lancet 2004; 363(9413):978–88.

[5] Khalil M, Teunissen CE, Otto M, Piehl F, Sormani MP, Gattringer T, et al. Neurofilaments as biomarkers in neurological disorders. Nature Reviews Neurology 2018; 14(10):577–89.

[6] Anand S, Hanson K. Disability-adjusted life years: a critical review. Journal of Health Economics 1997;16(6):685–702.

[7] Forero DA, López-León S, González-Giraldo Y, Dries DR, Pereira-Morales AJ, Jiménez KM, et al. *APOE* gene and neuropsychiatric disorders and endophenotypes:

a comprehensive review. American Journal of Medical Genetics Part B: Neuropsychiatric Genetics 2018;177(2):126–42.

[8] Meekins GD, So Y, Quan D. American Association of Neuromuscular & Electrodiagnostic Medicine evidenced-based review: use of surface electromyography in the diagnosis and study of neuromuscular disorders. Muscle & Nerve 2008;38(4):1219–24.

[9] Gasecka A, Siwik D, Gajewska M, Jaguszewski MJ, Mazurek T, Filipiak KJ, et al. Early biomarkers of neurodegenerative and neurovascular disorders in diabetes. Journal of Clinical Medicine 2020;9(9):2807.

[10] Ghajar J. Traumatic brain injury. The Lancet 2000;356(9233):923–9.

[11] Tamburin S, Smania N, Saltuari L, Hoemberg V, Sandrini G. New advances in neurorehabilitation. Frontiers in Neurology 2019;10:1090.

[12] Friederich A, Bernd T, de Witte L. Methods for the selection of assistive technology in neurological rehabilitation practice. Scandinavian Journal of Occupational Therapy 2010;17(4):308–18.

[13] Bamdad M, Zarshenas H, Auais MA. Application of BCI systems in neurorehabilitation: a scoping review. Disability and Rehabilitation: Assistive Technology 2015; 10(5):355–64.

[14] Alizadeh A, Dyck SM, Karimi-Abdolrezaee S. Traumatic spinal cord injury: an overview of pathophysiology, models and acute injury mechanisms. Frontiers in Neurology 2019;10:282.

[15] Syriopoulou-Delli CK, Eleni G. Effectiveness of different types of augmentative and alternative communication (AAC) in improving communication skills and in enhancing the vocabulary of children with ASD: a review. Review Journal of Autism and Developmental Disorders 2022;9(4):493–506.

[16] Arac A. Machine learning for 3D kinematic analysis of movements in neurorehabilitation. Current Neurology and Neuroscience Reports 2020;20(8):29.

[17] Sutton A, Trudeau N, Morford JP, Smith MM, Mascolo M-È. Expressive and receptive performance with graphic symbol sentences by individuals who use aided communication. Augmentative and Alternative Communication 2022;38(3):135–47.

[18] Peters B, Eddy B, Galvin-McLaughlin D, Betz G, Oken B, Fried-Oken M. A systematic review of research on augmentative and alternative communication brain- computer interface systems for individuals with disabilities. Frontiers in Human Neuroscience 2022;16:952380.

[19] Donaldson AL, Corbin E, McCoy J. "Everyone deserves AAC": preliminary study of the experiences of speaking autistic adults who use augmentative and alternative communication. Perspectives of the ASHA Special Interest Groups 2021;6(2): 315–26.

[20] Elsahar Y, Hu S, Bouazza-Marouf K, Kerr D, Mansor A. Augmentative and alternative communication (AAC) advances: a review of configurations for individuals with a speech disability. Sensors 2019;19(8):1911.

[21] Sarker IH. Machine learning: algorithms, real-world applications and research directions. SN Computer Science 2021;2(3):160.

[22] Doeltgen SH, Huckabee M-L. Swallowing neurorehabilitation: from the research laboratory to routine clinical application. Archives of Physical Medicine and Rehabilitation 2012;93(2):207–13.

[23] Esfahlani SS, Shirvani H, Butt J, Mirzaee I, Esfahlani KS. Machine learning role in clinical decision-making: neuro-rehabilitation video game. Expert Systems with Applications 2022;201:117165.

[24] Massetti T, da Silva TD, Crocetta TB, Guarnieri R, de Freitas BL, Bianchi Lopes P, et al. The clinical utility of virtual reality in neurorehabilitation: a systematic review. Journal of Central Nervous System Disease 2018;10:117957351881354.
[25] Khan F, Amatya B, Galea MP, Gonzenbach R, Kesselring J. Neurorehabilitation: applied neuroplasticity. Journal of Neurology 2017;264(3):603−15.
[26] Wang M, Cui Y, Wang X, Xiao S, Jiang J. Machine learning for networking: workflow, advances and opportunities. IEEE Network 2018;32(2):92−9.
[27] Herbst J. A machine learning approach to workflow management. 2000. p. 183−94.
[28] Frey S, Motawar B, Buchanan K, Kaufman C, Stevens P, Cirstea C, et al. Greater and more natural use of the upper limbs during everyday life by former amputees versus prosthesis users. Neurorehabilitation and Neural Repair 2022;36(3):227−38.
[29] Wheaton LA. Neurorehabilitation in upper limb amputation: understanding how neurophysiological changes can affect functional rehabilitation. Journal of NeuroEngineering and Rehabilitation 2017;14(1):41.
[30] Cusack WF, Thach S, Patterson R, Acker D, Kistenberg RS, Wheaton LA. Enhanced neurobehavioral outcomes of action observation prosthesis training. Neurorehabilitation and Neural Repair 2016;30(6):573−82.
[31] Wang N, Farhadi A, Rao R, Brunton B. AJILE movement prediction: multimodal deep learning for natural human neural recordings and video. Proceedings of the AAAI Conference on Artificial Intelligence 2018;32(1).
[32] Yousefi J, Hamilton-Wright A. Characterizing EMG data using machine- learning tools. Computers in Biology and Medicine 2014;51:1−13.
[33] Karlık B. Machine learning algorithms for characterization of EMG signals. International Journal of Information and Electronics Engineering 2014;4(3):189.
[34] Mohseni M, Shalchyan V, Jochumsen M, Niazi IK. Upper limb complex movements decoding from pre-movement EEG signals using wavelet common spatial patterns. Computer Methods and Programs in Biomedicine 2020;183:105076.
[35] Edwards AL, Dawson MR, Hebert JS, Sherstan C, Sutton RS, Chan KM, et al. Application of real-time machine learning to myoelectric prosthesis control. Prosthetics and Orthotics International 2016;40(5):573−81.
[36] Zanker M, Rook L, Jannach D. Measuring the impact of online personalisation: past, present and future. International Journal of Human-Computer Studies 2019;131: 160−8.
[37] Xu Z, Zhou X, Li S. Deep recurrent neural networks based obstacle avoidance control for redundant manipulators. Frontiers in Neurorobotics 2019;13:47.
[38] Hutabarat D, Rivai M, Purwanto D, Hutomo H. Lidar-based obstacle avoidance for the autonomous mobile robot. 2019 12th International Conference on Information & Communication Technology and System (ICTS). 2019. p. 197−202.
[39] Farag KKA, Shehata HH, El-Batsh HM. Mobile robot obstacle avoidance based on neural network with a standardization technique. Journal of Robotics 2021;2021: 1−14.
[40] Cuaya-Simbro G, Perez-Sanpablo A-I, Morales E-F, Quiñones Uriostegui I, Nuñez-Carrera L. Comparing machine learning methods to improve fall risk detection in elderly with osteoporosis from balance data. Journal of Healthcare Engineering 2021;2021:1−11.
[41] Bargiotas I, Wang D, Mantilla J, Quijoux F, Moreau A, Vidal C, et al. Preventing falls: the use of machine learning for the prediction of future falls in individuals without history of fall. Journal of Neurology 2022;270:618−31.

[42] Virgilio González CD, Sossa Azuela JH, Antelis JM, Falcón LE. Spiking Neural Networks applied to the classification of motor tasks in EEG signals. Neural Networks 2020;122:130–43.

[43] Sennott SC, Akagi L, Lee M, Rhodes A. AAC and artificial intelligence (AI). Topics in Language Disorders 2019;39(4):389–403.

[44] Neamtu R, Camara A, Pereira C, Ferreira R. Using artificial intelligence for augmentative alternative communication for children with disabilities. 2019. p. 234–43.

[45] Langton-Frost N, Brodsky MB. Speech-language pathology approaches to neurorehabilitation in acute care during COVID-19: capitalizing on neuroplasticity. PM&R 2022;14(2):217–26.

[46] Nadkarni PM, Ohno-Machado L, Chapman WW. Natural language processing: an introduction. Journal of the American Medical Informatics Association 2011;18(5): 544–51.

[47] Chowdhary KR. Natural language processing. In: Fundamentals of artificial intelligence. New Delhi: Springer India; 2020. p. 603–49.

[48] Ugartemendia A, Rosquete D, Gil JJ, Diaz I, Borro D. Machine learning for active gravity compensation in Robotics: application to neurological rehabilitation systems. IEEE Robotics and Automation Magazine 2020;27(2):78–86.

[49] Tozlu C, Edwards D, Boes A, Labar D, Tsagaris KZ, Silverstein J, et al. Machine learning methods predict individual upper-limb motor impairment following therapy in chronic stroke. Neurorehabilitation and Neural Repair 2020;34(5):428–39.

[50] Liu A, Lee S, Chen X, McKeown MJ, Wang ZJ. Galvanic vestibular stimulation: data analysis and applications in neurorehabilitation. IEEE Signal Processing Magazine 2021;38(4):54–64.

[51] Kurkin S, Chholak P, Maksimenko V, Pisarchik A. Machine learning approaches for classification of imaginary movement type by MEG data for neurorehabilitation. In: 2019 3rd School on Dynamics of Complex Networks and their Application in Intellectual Robotics (DCNAIR); 2019. p. 106–8.

[52] Mazzoleni S, Turchetti G, Palla I, Posteraro F, Dario P. Acceptability of robotic technology in neuro-rehabilitation: preliminary results on chronic stroke patients. Computer Methods and Programs in Biomedicine 2014;116(2):116–22.

[53] Simonetti D, Zollo L, Papaleo E, Carpino G, Guglielmelli E. Multimodal adaptive interfaces for 3D robot-mediated upper limb neuro-rehabilitation: an overview of bio-cooperative systems. Robotics and Autonomous Systems 2017;90:86–96.

[54] Soekadar SR, Birbaumer N, Slutzky MW, Cohen LG. Brain–machine interfaces in neurorehabilitation of stroke. Neurobiology of Disease 2015;83:172–9.

[55] Kubota A, Riek LD. Methods for robot behavior adaptation for cognitive neurorehabilitation. Annual Review of Control, Robotics, and Autonomous Systems 2022;5(1):109–35.

[56] Calvo I, Tropea P, Viganò M, Scialla M, Cavalcante AB, Grajzer M, et al. Evaluation of an automatic speech recognition platform for dysarthric speech. Folia Phoniatrica et Logopaedica 2021;73(5):432–41.

[57] Viroli C, McLachlan GJ. Deep Gaussian mixture models. Statistics and Computing 2019;29(1):43–51.

[58] McLachlan GJ, Rathnayake S. On the number of components in a Gaussian mixture model. Wiley Interdisciplinary Reviews: Data Mining and Knowledge Discovery 2014;4(5):341–55.

[59] Fernandes FG, de Santana EM. Machine learning applied to neuroreabilitation: a systematic mapping. Journal on Advances in Theoretical and Applied Informatics 2020; 5(1).
[60] Avutu SR, Paul S. Artificial intelligence algorithms for healthcare and neurorehabilitation engineering. 2022. p. 103–18.
[61] Hamid H, Naseer N, Nazeer H, Khan MJ, Khan RA, Shahbaz Khan U. Analyzing classification performance of fNIRS-BCI for gait rehabilitation using deep neural networks. Sensors 2022;22(5):1932.
[62] Hernández LG, Antelis JM. A comparison of deep neural network algorithms for recognition of EEG motor imagery signals. 2018. p. 126–34.
[63] Lin AS, Pratik S, Ota J, Rawat TS, Huang T-H, Hsu C-L, et al. A process-aware memory compact-device model using long-short term memory. IEEE Access 2021;9: 3126–39.
[64] Latif SD. Concrete compressive strength prediction modeling utilizing deep learning long short-term memory algorithm for a sustainable environment. Environmental Science and Pollution Research 2021;28(23):30294–302.
[65] Montazerin M, Zabihi S, Rahimian E, Mohammadi A, Naderkhani F. ViT- HGR: vision transformer-based hand gesture recognition from high density surface EMG signals. 2022.
[66] Wang Z, Ma Y, Liu Z, Tang J. R-transformer: recurrent neural network enhanced transformer. 2019.
[67] Tayeb Z, Fedjaev J, Ghaboosi N, Richter C, Everding L, Qu X, et al. Validating deep neural networks for online decoding of motor imagery movements from EEG signals. Sensors 2019;19(1):210.
[68] Karácsony T, Hansen JP, Iversen HK, Puthusserypady S. Brain computer interface for neuro-rehabilitation with deep learning classification and virtual reality feedback. Proceedings of the 10th Augmented Human International Conference 2019. 2019. p. 1–8.
[69] Sun S, Huang R. An adaptive k-nearest neighbor algorithm. 2010 Seventh International Conference on Fuzzy Systems and Knowledge Discovery. 2010. p. 91–4.
[70] Cunningham P, Delany SJ. K-nearest neighbour classifiers - a tutorial. ACM Computing Surveys 2022;54(6):1–25.
[71] Thanh Noi P, Kappas M. Comparison of random forest, k-nearest neighbor, and support vector machine classifiers for land cover classification using sentinel-2 imagery. Sensors 2017;18(2):18.
[72] Federici S, Meloni F, Bracalenti M, de Filippis ML. The effectiveness of powered, active lower limb exoskeletons in neurorehabilitation: a systematic review. NeuroRehabilitation 2015;37(3):321–40.
[73] Huang S-C, Guerrieri S, Dalla Costa G, Pisa M, Leccabue G, Gregoris L, et al. Intensive neurorehabilitation and gait improvement in progressive multiple sclerosis: clinical, kinematic and electromyographic analysis. Brain Sciences 2022;12(2):258.
[74] Bui KD, Johnson MJ. Designing robot-assisted neurorehabilitation strategies for people with both HIV and stroke. Journal of NeuroEngineering and Rehabilitation 2018; 15(1):75.
[75] Poewe W, Seppi K, Tanner CM, Halliday GM, Brundin P, Volkmann J, et al. Parkinson disease. Nature Reviews Disease Primers 2017;3(1):17013.
[76] Armstrong MJ, Okun MS. Diagnosis and treatment of Parkinson disease. JAMA 2020; 323(6):548.

[77] Giuffrida JP, Lerner A, Steiner R, Daly J. Upper-extremity stroke therapy task discrimination using motion sensors and electromyography. IEEE Transactions on Neural Systems and Rehabilitation Engineering 2008;16(1):82–90.

[78] Haque F, Nancel M, Vogel D. Myopoint. Proceedings of the 33rd Annual ACM Conference on Human Factors in Computing Systems. 2015. p. 3653–6.

[79] Paulo J, Garrote L, Premebida C, Asvadi A, Almeida D, Lopes A, et al. An innovative robotic walker for mobility assistance and lower limbs rehabilitation. In: 2017 IEEE 5th Portuguese Meeting on Bioengineering (ENBENG); 2017. p. 1–4.

[80] Lin C-T, Chiang H-H, Lee T-T. A practical fuzzy controller with Q-learning approach for the path tracking of a walking-aid robot. The SICE annual conference. 2013; 2013. p. 888–93.

[81] Hartigan JA, Wong MA. Algorithm AS 136: a K-means clustering algorithm. Applied Statistics 1979;28(1):100.

[82] Maitre NL. Neurorehabilitation after neonatal intensive care: evidence and challenges. Archives of Disease in Childhood - Fetal and Neonatal Edition 2015;100(6):F534–40.

[83] Smouha E. Inner ear disorders. NeuroRehabilitation 2013;32(3):455–62.

[84] Aisen ML, Kerkovich D, Mast J, Mulroy S, Wren TA, Kay RM, et al. Cerebral palsy: clinical care and neurological rehabilitation. The Lancet Neurology 2011;10(9): 844–52.

[85] Oguntosin VW, Mori Y, Kim H, Nasuto SJ, Kawamura S, Hayashi Y. Design and validation of exoskeleton actuated by soft modules toward neurorehabilitation—vision-based control for precise reaching motion of upper limb. Frontiers in Neuroscience 2017;11:352.

[86] Calabrò RS, Cerasa A, Ciancarelli I, Pignolo L, Tonin P, Iosa M, et al. The arrival of the metaverse in neurorehabilitation: fact, fake or vision? Biomedicines 2022;10(10): 2602.

[87] Wang M, Reid D. Virtual reality in pediatric neurorehabilitation: attention deficit hyperactivity disorder, autism and cerebral palsy. Neuroepidemiology 2011;36(1):2–18.

[88] Leopold A, Lourie A, Petras H, Elias E. The use of assistive technology for cognition to support the performance of daily activities for individuals with cognitive disabilities due to traumatic brain injury: the current state of the research. NeuroRehabilitation 2015;37(3):359–78.

[89] Stock SE, Davies DK, Wehmeyer ML, Lachapelle Y. Emerging new practices in technology to support independent community access for people with intellectual and cognitive disabilities. NeuroRehabilitation 2011;28(3):261–9.

[90] Bos HR, Babbage DR, Leathem JM. Efficacy of memory aids after traumatic brain injury: a single case series. NeuroRehabilitation 2017;41(2):463–81.

[91] Särkämö T. Cognitive, emotional, and neural benefits of musical leisure activities in aging and neurological rehabilitation: a critical review. Annals of Physical and Rehabilitation Medicine 2018;61(6):414–8.

[92] Kapur N, Glisky EL, Wilson BA. Technological memory aids for people with memory deficits. Neuropsychological Rehabilitation 2004;14(1–2):41–60.

[93] Kwakkel G, Kollen BJ, Krebs HI. Effects of robot-assisted therapy on upper limb recovery after stroke: a systematic review. Neurorehabilitation and Neural Repair 2008; 22(2):111–21.

[94] Altenmüller E, Schlaug G. Neurologic music therapy: the beneficial effects of music making on neurorehabilitation. Acoustical Science and Technology 2013;34(1):5–12.

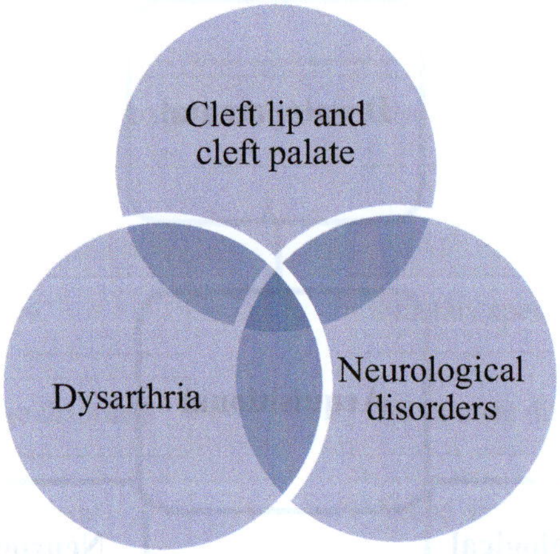

FIGURE 7.3

Hypernasality.

cases such as allergies, cleft palate and large tonsils, and medical and surgical interventions may be required.

1.1.3 Voice disorders

It can be defined as disorders observed in the structure, quality, and tone of the voice, which are phonological features. The most common symptoms in people with voice disorders are pain around the throat and the vocal cord and shortness of breath. The voice is thinner or thicker than normal. Voice disorders can manifest due to reasons such as a nodule, cyst, edema, vocal cord paralysis, excessive smoking, alcohol use, and respiratory tract infection. The term "voice disorder" embodies many different conditions with signs and symptoms that can present in isolation or combination with each other. It requires the expertise of multiple professionals such as otolaryngologists (ear, nose, and throat), Speech–language pathologists (SLP), voice scientists, psychologists, singing teachers, and vocal coaches (Fig. 7.4) [5].

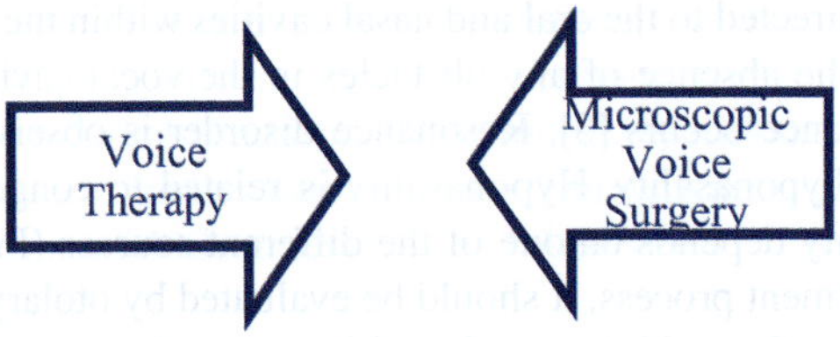

FIGURE 7.4

Voice disorders treatment methods.

1.1.4 Motor speech disorders

Speech is a physical and mental activity that we perform with the speech organs. Motor speech disorder is a neuromotor problem that affects one or more of the components of breathing, vocalization, resonance, articulation, and prosody. There are two types of motor speech disorders, namely dysarthria and apraxia. Dysarthria is a motor speech disorder that occurs in the muscles that control the ability to speak due to damage to the central nervous system and peripheral nervous system. Dysarthria is a motor speech disorder. The sources of neuromotor problems that cause dysarthria are vascular, traumatic, infectious, neoplastic, and metabolic. Dysarthria in adults occurs due to brain injuries such as brain attack (stroke) and brain trauma. Neuromotor speech disorders observed in childhood are defined as developmental dysarthria. Speech apraxia, on the other hand, is a motor disorder related to the programming of speech in adulthood and childhood.

In the treatment of speech disorders such as apraxia and dysarthria, the treatment process is long term due to its multisource nature and it is recommended to begin as soon as possible. Especially, in dysarthria, a calendar is determined for the underlying cause. The auditory-perceptual approach is currently the "gold standard" for the diagnostic assessment of motor speech disorders in clinical practice. It is also used for assessing the severity and changes in voice and speech over time [6]. In the planning of therapy processes, linguistic skills should be given importance in addition to motor skills. The goals of speech and language therapy for children with motor speech disorders include increasing speech production and decisiveness. It also improves general communication and language skills using alternative communication methods [7].

1.1.5 Language disorders

Language disorders can be expressed with the concepts of delayed language or developmental language disorder, which do not have a specific known cause. In addition, it covers language disorders that develop due to a specific cause such as mental disabilities, neurological problems, and hearing impairment. Language disorder and speech disorder are two separate disorders that should not be confused with each other. If a child has difficulty in understanding oral communication or communicating his feelings and thoughts verbally at an adequate level, then this is a language disorder. On the other hand, if a child has difficulty producing speech sounds correctly or fluently, then this is a speech disorder.

Language development in individuals begins with listening and imitation skills. The hearing system of the baby records everything related to language during pregnancy and through the organs that enable speech after birth. The stages of language development in the baby are divided into two stages: prespeech and postspeech. Before the age of 2, the child performs grammar using signs such as gestures—facial expressions along with words.

Basically, the first factor related to language disorders is medical and the second is social-environmental factors [8]. As can be seen in Fig. 7.5, individuals with a family history of language disorder are more likely to be at risk due to genetic

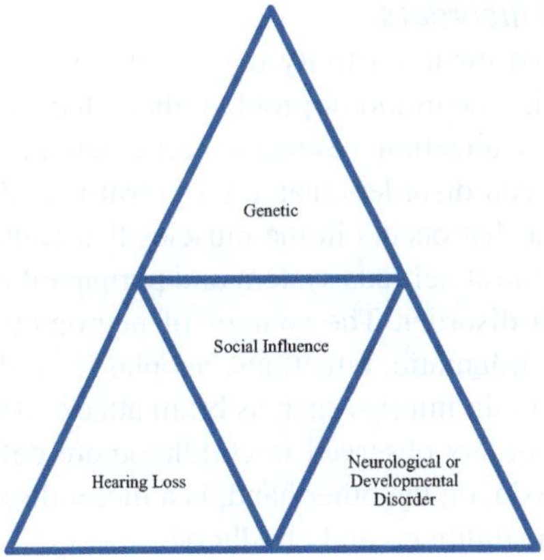

FIGURE 7.5

Factors that may cause language disorder.

characteristics. At the same time, the family environment in which the individual was born is very important. Babies who have parents who talk to them and communicate with them acquire faster and healthier speech skills. In addition to these, hearing loss also hinders the understanding of verbal language. It is observed that language skills are affected in individuals with neurological problems. In a study, Toseeb, Oginni, and Dale [9] has added to the existing evidence that common genetic influences the co-occurrence of developmental language disorder and psychopathology. It has presented preliminary evidence that genetic's influence on internal problems is stronger in young people with a developmental language disorder. Individuals who experience language disorders communicate using fewer sentences, unlike others. The treatment process is applied to encourage the use of sentences. These problems can be corrected with speech and language therapy.

1.1.6 Acquired language disorders

It is a language disorder that occurs due to damage to the areas of the brain responsible for language. Language disorder, which usually appears suddenly as a result of brain injuries such as stroke or head trauma, is caused by damage to the areas of the brain responsible for language. It appears in the form of aphasia. However, head injuries, infections, or head traumas are the main causes of acquired language disorder.

Aphasia is an acquired language disorder that occurs due to brain damage. As a result of the damage caused to the speech centers in the brain, it can be observed in the form of partial or complete loss of skills such as speaking, understanding, naming, perception, reading, or writing. Aphasia is not a mental disability but they cannot perform comprehension and explanation actions. Changing the standard language pattern, changing the sounds in the word, creating meaningless sounds and word groups, and deciphering sentences without paying attention to syntax are among the most obvious symptoms of acquired language disorder.

In some cases, it is possible to wait for recovery without applying a treatment process to individuals with aphasia. Improvements that occur spontaneously in this way are seen in cases where the blood flow to the brain stops temporarily, but then returns to normal order. In all cases, the recovery and return of the tongue may not be so fast. Therefore, sometimes cases of permanent aphasia are also mentioned. In such cases, assistance from language and speech therapy should be sought. The speech and language therapist recreates language skills and teaches communication. During this process, the patient may need psychological support. The treatment process can be carried out under the guidance of a psychologist. Although aphasia is mostly seen in adults, aphasia acquired in childhood is a rare and controversial topic. The knowledge about childhood aphasia remains limited [10].

1.1.7 Speech sound disorders

Disorders caused by structural or motor-based problems that occur in the speech organs are called speech sound disorders. Speech sound disorder is a condition that impairs the intelligibility of speech or prevents the transmission of a message, which is an important element in communication.

Symptoms begin at the early stage of language development. Congenital syndromes, hearing loss, and neurological conditions cause speech sound disorders. From a psycholinguistic perspective, children's speech consists of speech input processing, speech output processing, and speech production stages. Children's speech difficulties can be caused by impairment in one or more speech-processing areas [11]. Specialist speech–language therapists should be consulted at an early stage for speech sound disorder. Incorrect sounds should be detected and an appropriate treatment should be performed for the target sound. Activities based on the correct pronunciation of the sound under the supervision of a therapist will help to use that sound correctly in everyday life.

According to American Speech–Language–Hearing Association (ASHA) data, a number of statistics for language and speech disorders [12] are given.

1. "Nearly 1 in 12 (7.7 percent) of U.S. children with age 3–17 has had a disorder related to voice, speech, language or swallowing in the past 12 months."
2. "Among children who have a voice, speech, language or swallowing disorder, 34 percent of those with ages 3–10 have multiple communication or swallowing disorders while 25.4 percent of those with ages 11–17 have multiple disorders."
3. "Boys with ages 3–17 are more likely than girls to have a voice, speech, language, or swallowing disorder (9.6 percent compared to 5.7 percent)."

The mental disability status, which is considered among the factors that may cause language and speech disorders, is a subject that needs to be evaluated. Individuals with mild intellectual disabilities make articulation errors such as skipping sounds, adding sounds, or mispronouncing sounds [13]. In addition, they have difficulty in understanding oral communication and cannot respond and form sentences due to the lack of vocabulary. It is also found that they more often use the wrong sentence.

Even if they have the skills to communicate with the individuals around them, they may not be able to fully express themselves to everyone because their vocabulary is not sufficient. Articulation disorder is frequently observed in these individuals. When the ratio of Developmental Intellectual Disability to the population among the countries of the world up to 2019 is examined (The population by Idiopathic "Developmental Intellectual Disability"), the highest rate is observed in Syria, Sudan, Yemen, Somalia, Afghanistan, and India, followed by Morocco, Libya, Nigeria, Republic of Congo, Egypt, Mozambique, Papua New Guinea, and on the other hand, a very small percent was found in Denmark and South Korea [14].

1.1.7.1 Deep learning and machine learning in AI

The system of machines that develop methods that imitate human thinking ability is called AI. AI is the process of giving human thinking ability to machines. In this context, it covers machine learning. Deep learning, on the other hand, is a subset of machine learning in which artificial neural networks and algorithms inspired by the human brain learn from data. The main difference between these two is based on the decoding process in the system. While machine learning always needs structured data, deep learning networks work at the level of artificial neural networks (Fig. 7.6).

Machine learning refers broadly to the process of fitting predictive models to data or of identifying informative groupings within data [15]. Machine learning tries to create a skill close to humans using machine learning calculations. ML has emerged from the automated process of computers to provide a solution for the problem and continues to operate in many areas such as voice recognition systems, intelligent

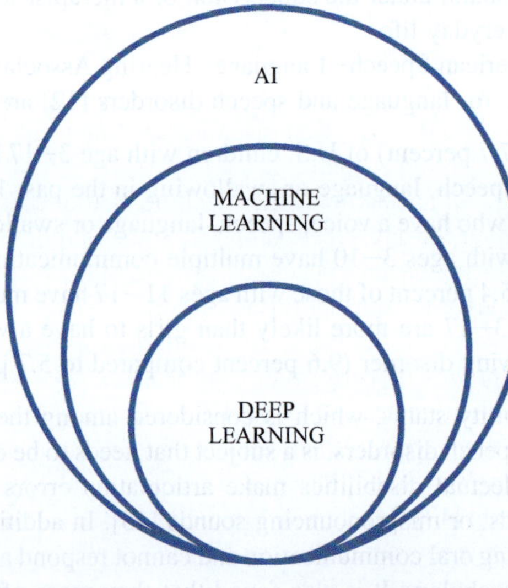

FIGURE 7.6

The relation with AI, ML, and DP.

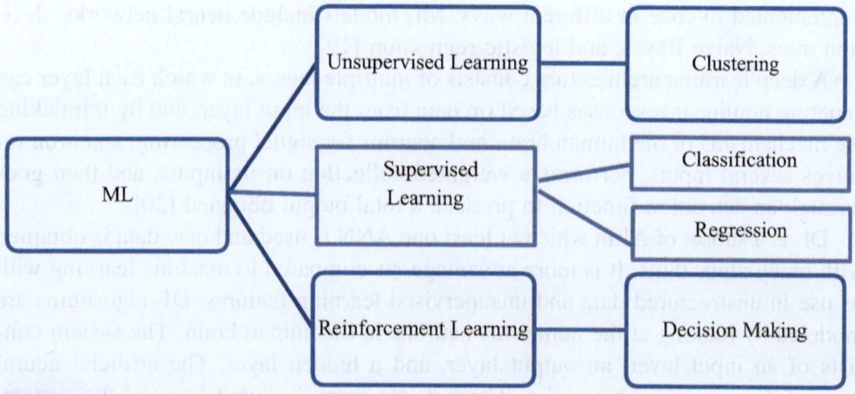

FIGURE 7.7
Types of ML.

assistants, and self-driving cars [16]. The ability to reason with sound and image processing is related to AI. Deep learning forms the basis of both AI and machine learning. DL can do self-learning by processing computer algorithms. It works with artificial neural networks (ANNs) that mimic the process of thinking procedure of humans. DL uses a huge amount of data and is a group of machine learning consisting of multilayered ANNs.

ML uses algorithms to make decisions based on classified data. DL structures algorithms in layers to create an ANN that can learn on its own and make realistic decisions (Fig. 7.7).

There are four types of ML algorithms such as supervised, semisupervised, unsupervised, and reinforcement. ML involves a set of algorithms that work on big data. Supervised machine learning has a supervisory mechanism. First, the machine is trained with inputs and outputs and then the machine is asked to estimate the output using the test dataset. Supervised machine learning is divided into two types, namely classification and regression.

In unsupervised machine learning, the machine is trained using an unlabeled dataset without the need for any supervision. Then the output is estimated without an audit. The purpose of the unsupervised learning algorithm is to classify data sets into different categories. Unsupervised learning is classified into two types: clustering and association. Reinforcement learning is a feedback-based type in which an AI tool improves its performance by learning from past experiences. It is inspired by behaviorism instead of labeled data. Reinforcement learning is categorized basically into two types of methods/algorithms: positive and negative RL.

Supervised learning should be preferred if we have a smaller amount of clearly labeled data for training [17]. ML is widely used in both research and industrial applications. With the huge success of DL and NNs, their influence and possible aftereffects should be taken seriously [18]. ML encompasses a number of models that are implemented in code in different ways. ML models include neural networks, decision trees, Naive Bayes, and logistic regression [19].

A deep learning architecture consists of multiple layers, in which each layer can generate nonlinear responses based on data from the input layer, and by mimicking the mechanisms of the human brain and neurons for signal processing, a neuron receives several inputs, performs a weighted collection on its inputs, and then goes through an activation function to produce a total output obtained [20].

DL is a subset of AI in which at least one ANN is used and new data is obtained with many algorithms. It is more advantageous compared to machine learning with its use in unstructured data and unsupervised learning features. DL algorithms are modeled by looking at the numerous neurons in the human brain. The system consists of an input layer, an output layer, and a hidden layer. The artificial neural network has several input nodes. These nodes form the input layer of the system. The input layer processes the data and transmits it to the layers deep inside the neural network. The output layer consists of nodes that give output data.

As can be seen in Table 7.1, ML also needs human intervention to learn more from data. DL can retrieve unstructured data in its raw form and automatically identify clusters. While conventional methods do not focus too much on accuracy, ML aims to improve accuracy. In light of this comparison, AI in the field of health is used in individual medicine, in findings obtained from different diseases, and in the discovery of new treatments. Medical fields where AI is expected to be easily applied to functional use include genomic medicine, diagnostic imaging support, diagnosis and treatment support, and drug discovery [21].

In general terms, there are two types of AI methods in healthcare: ML and DL [22]. ML and DL have made great advances and the success of AI in the medical field has been highlighted by significant work in medical AI applications [23]. AI can view large amounts of data from health institutions and use it in clinical studies and forecasts. With DL, data is analyzed and interpreted by computers with the help of extended information. The use of AI in the health sector serves many stakeholders. Researchers conducting clinical trials, teams consisting of data managers can be shown as examples. It can be used to develop a personalized health plan, predict diseases, and accelerate diagnosis.

Table 7.1 Comparison of ML and DL.

DL	ML
A subset of ML	A subset of AI
Trains on larger data	Trains on smaller data
Learns from environment	Needs humans to learn
Trains model	Trains system
Makes nonlinear correlations	Makes linear correlations

2. Research methodology

This study was conducted as a literature review. Literature review is the examination of previously published studies on a topic under a category [24]. In this method, the goal is to put forward scientific evidence based on research written on the subject.

2.1 Research questions

The research questions addressed by the study are as follows.

Q1. What is a language and speech disorder as a symptom of a nervous system disorder?
Q2. Why are artificial intelligence, machine learning, and deep learning used in the field of healthcare?
Q3. What is the success rate of DL and ML method in patients with speech and language disorders?

2.2 Data collection and data analysis process

The imprint information obtained from each article included in the research is as follows.

- Scope
- Country
- Year

The collection of data consisted of the previously mentioned processes. The sections related to the content information of the surveys are shown in tables. Table 7.2 consists of the following information. Then there is a detailed demonstration of each research.

- What is the scope of the investigations?
- What is the research topic?
- Country of origin of the research?
- Year in which research was conducted?

3. DL and ML methods for patients with SLD

Kuo et al. [25] have developed a system that analyzes acoustic data for the diagnosis and classification of voice disorders in children with the help of a computer. According to this work, data were collected from 90 children aged 3–6 with pathological articulatory features. Three neural network models were used for the classification of speech samples. The experiments examined the availability of four different categories of Chinese phrases and characters. The objective of this work is to use neural networks to classify speech sound disorders (SSD) in both binary and multilevel

Table 7.2 Research in DL and ML methods used in patients with SLD.

Research code	Year	Country	Scope
RC1	2019	India	ML in the process of understanding voice disorder
RC2	2022	India	Speech disorder and ML
RC3	2022	France	Stuttering and ML/DL
RC4	2019	Greece	Language disorders and AI
RC5	2020	India	Speech disorders and SVM
RC6	2022	India	Stuttering and DL
RC7	2022	India	Speech disorders and ML/DL
RC8	2022	Malaysia	Cleft lip/Palate and ML
RC9	2022	United States	Adult dysarthritic speech, children with cleft lip/ palate accent problems and DL
RC10	2022	China/ Malaysia/ Iran	Parkinson's disease and cleft lip and palate of pathological speech diagnosis and DCNN
RC11	2022	China	Speech disorders and DL
RC12	2022	Australia	Language disorders and ML
RC13	2022	India	Speech technology and DL
RC14	2022	China	Hypernasality estimation in cleft palate speech and ML
RC15	2022	India	Orthognathic surgery and ML
RC16	2022	India	Dysarthria and DL
RC17	2022	New Zealand	Dysarthric intelligibility and DL technology
RC18	2022	Switzerland	Dysarthric speech and DL

categories. The features of convolutional neural network (CNN) models were used. Four common types of SSD errors have been classified using monosyllabic speech examples and neural network models.

Janbakhshi and Kodrasi [26] have worked on dysarthric speech detection approaches based on deep learning. DL methods learn discriminatory representations by processing the broad signal spectrum and separating the phase spectrum. In their article, the applicability of short-time fourier transform (STFT) phase representations in dysarthric speech detection has been investigated. A double-input CNN was used. It was found that it is important to take into account the phase information for a dysarthric speech disorder. In the research of Hall, Huang, and Shahamiri [27], it is noted that recent developments in the deep learning method offer a cost-effective, accessible, and less subjective way to assess patients with dysarthria. More than 13 experiments were conducted in the research, It was evaluated to determine whether dysarthric intelligibility assessment models were generalized to new speakers.

Joshy and Rajan [28] have stated that it is easy to determine the severity level of dysarthria. It is used to give an idea on the patient's recovery and to help automatic dysarthria speech recognition systems. In this research, a comparative study on the classification of dysarthria severity levels using different deep learning techniques and acoustic features is presented. Mohaideen et al. [29] provided an overview on the applications of AI and ML algorithms in orthognathic surgery. A total of 19 studies were examined. After applying the inclusion and exclusion criteria, a total of 19 studies were included in the final review. As a result, it was found that AI reduces the workload of the clinician by eliminating tedious registration procedures.

Song et al. [30] stated that hypernasality prediction is very important in the diagnosis of cleft palate due to hypernasality prediction and an abnormality in human speech resonance. Methods for hypernasality prediction have been proposed in this research on performing acoustic analysis based on low-resource cleft palate datasets using statistical or neural network-based features.

Deepa and Khilar [31] have stated that deep learning technology is used for disease detection, monitoring, analysis, and new-generation companion understanding in speech disorders. Gasparini et al. [32] reported that the majority of children from the samples have shown improvement after receiving the therapy. An improvement of 73% in sensitivity and 77% in specificity is reported in the article. Murugan et al. [33] used a classification algorithm with CNN to analyze the stuttered audio signal. Arpitha et al. [34] have proposed another work in which the aim is to improve the effective communication skills of people who have many speech disorders due to stroke. Speech acoustic signals were obtained from the speaker, which were converted and given as input to the microphone.

Gkeka, Agorastou, and Drigas [35] have emphasized that AI techniques and applications should be used for learning and teaching language disorders and deficits. It has been emphasized that along with language disorders, it is necessary to develop prediction systems in the areas of written language, oral language, and phoneme pronunciation and communication. Sheikh et al. [36] emphasized that stuttering should be evaluated together with fields such as pathology, psychology, and acoustics. Anthony, Patil, and Basavaiah [37] have stated that speech disorders can occur in multiple forms in people with different symptoms and various tools should be developed to detect these changes. Advances in speech signal processing and natural language processing through DNN have been given much emphasis in this work.

Hegde et al. [38] used a machine learning method to identify voice disorders in their research. Yao, Chi, and Khishe [39] used a whale optimization algorithm to automatically select the best architecture for DCNNs for the investigation of pathological sound definition. Mathad et al. [40] have used CNN in their research on adult dysarthric speech, speech of children with cleft lip/palate, and individuals with accentuated speech disorders. Huqh et al. [41] have provided a systematic review to investigate the current clinical applications of AI/ML techniques in the diagnosis and treatment of children with cleft lip and palate syndrome. Evangeline [42] has also done a systematic review to analyze innovative AI systems for abnormal speech situations. The authors stated that personalized methods have been developed with AI.

References

[1] Chenausky KV, Tager-Flusberg H. The importance of deep speech phenotyping for neurodevelopmental and genetic disorders: a conceptual review. Journal of Neurodevelopmental Disorders 2022;14(1):1−14.

[2] Laiho A, Elovaara H, Kaisamatti K, Luhtalampi K, Talaskivi L, Pohja S, et al. Stuttering interventions for children, adolescents, and adults: a systematic review as a part of clinical guidelines. Journal of Communication Disorders 2022:106242.

[3] Aksu İ, Topbaş S. 8-11 yaş Çocuklarının Nazometrik Değerleri. Dil Konuşma ve Yutma Araştırmaları Dergisi 2020;3(2):130−52.

[4] Liu Y, Lee SAS, Chen W. The correlation between perceptual ratings and nasalance scores in resonance disorders: a systematic review. Journal of Speech, Language, and Hearing Research 2022:1−20.

[5] Payten CL, Chiapello G, Weir KA, Madill CJ. Frameworks, terminology and Definitions used for the classification of voice disorders: a scoping review. Journal of Voice 2022:1−18.

[6] Pernon M, Assal F, Kodrasi I, Laganaro M. Perceptual classification of motor speech disorders: the role of severity, speech task, and listener's expertise. Journal of Speech, Language, and Hearing Research 2022;65(8):2727−47.

[7] Çağlar NA, Erbaş AN. Vaka Çalışması: Çocukluk Çağı Konuşma Apraksisi Olan bir Çocukta Dil ve Konuşma Müdahalesinin Etkililiğinin İncelenmesi. Ergoterapi ve Rehabilitasyon Dergisi 2022;10(3):127−32.

[8] Siregar I. The effectiveness of multisensory stimulation therapy in people with specific language disorder. Budapest International Research and Critics Institute (BIRCI-Journal): Humanities and Social Sciences 2022;5(1):5315−25.

[9] Toseeb U, Oginni OA, Dale PS. Developmental language disorder and psychopathology: Disentangling shared genetic and environmental influences. Journal of Learning Disabilities 2022;55(3):185−99.

[10] Gilardone G, Viganò M, Cassinelli D, Fumagalli FM, Calvo I, Gilardone M, et al. Post-stroke acquired childhood aphasia. A scoping review. Child Neuropsychology 2022: 1−26.

[11] Berti LC, de Assis MF, Cremasco E, Cardoso ACV. Speech production and speech perception in children with speech sound disorder. Clinical Linguistics and Phonetics 2022;36(2−3):183−202.

[12] URL1. https://www.asha.org/about/press-room/quick-facts/#:~:text=5%25%20to%20 10%25%20of%20Americans,children%20have%20noticeable%20speech%20disorders.

[13] Ünsal S, Uçak MM, Fatih BAL, SARIGÜL AY, Cengiz DU. Zihinsel engelli 3-7 yaş arasındaki çocuklarda sesletim (artikülasyon) bozukluklarının değerlendirilmesi. Güncel Pediatri 2017;15(3):17−25.

[14] The population by idiopathic "developmental intellectual disability." https:// ourworldindata.org/neurodevelopmental-disorders.

[15] Greener JG, Kandathil SM, Moffat L, Jones DT. A guide to machine learning for biologists. Nature Reviews Molecular Cell Biology 2022;23(1):40−55.

[16] Martins RM, Gresse Von Wangenheim C. Findings on teaching machine learning in high school: a ten-year systematic literature review. Informatics in Education 2022. https://doi.org/10.15388/infedu.2023.18.

[17] Mahesh B. Machine learning algorithms-a review. International Journal of Science and Research 2020;9:381−6.
[18] Tjoa E, Guan C. A survey on explainable artificial intelligence (xai): toward medical xai. IEEE Transactions on Neural Networks and Learning Systems 2020;32(11): 4793−813.
[19] Burrell J. How the machine 'thinks': understanding opacity in machine learning algorithms. Big Data & Society 2016;3(1).
[20] Chahal A, Gulia P. Machine learning and deep learning. International Journal of Innovative Technology and Exploring Engineering 2019;8(12):4910−4.
[21] Tsuneki M. Deep learning models in medical image analysis. Journal of Oral Biosciences 2022:312−20.
[22] Loh HW, Ooi CP, Seoni S, Barua PD, Molinari F, Acharya UR. Application of explainable artificial intelligence for healthcare: a systematic review of the last decade (2011−2022). Computer Methods and Programs in Biomedicine 2022:107161.
[23] Zhang Y, Weng Y, Lund J. Applications of explainable artificial intelligence in diagnosis and surgery. Diagnostics 2022;12(2):237. 1-18.
[24] Köroğlu SA. Literatür taramasi üzerine notlar ve bir tarama tekniği. GiDB Dergi 2015;(01):61−9.
[25] Kuo YM, Ruan SJ, Chen YC, Tu YW. Deep learning-based automated classification of Chinese speech sound disorders. arXiv 2022. preprint arXiv:2205.11748.
[26] Janbakhshi P, Kodrasi I. Experimental investigation on STFT phase representations for deep learning-based dysarthric speech detection. In: ICASSP 2022-2022 IEEE international conference on acoustics, speech and signal processing (ICASSP). IEEE; May 2022. p. 6477−81.
[27] Hall K, Huang A, Shahamiri SR. An investigation to identify optimal setup for automated assessment of dysarthric intelligibility using deep learning technologies. Cognitive Computation 2022:1−13.
[28] Joshy AA, Rajan R. Automated dysarthria severity classification: a study on acoustic features and deep learning techniques. IEEE Transactions on Neural Systems and Rehabilitation Engineering 2022;30:1147−57.
[29] Mohaideen K, Negi A, Verma DK, Kumar N, Sennimalai K, Negi A. Applications of artificial intelligence and machine learning in orthognathic surgery: a scoping review. Journal of Stomatology, Oral and Maxillofacial Surgery 2022:e962−72.
[30] Song K, Wan T, Wang B, Jiang H, Qiu L, Xu J, et al. Improving hypernasality estimation with automatic speech recognition in cleft palate speech. arXiv 2022. preprint arXiv:2208.05122.
[31] Deepa P, Khilar R. Speech technology in healthcare. Measurement: Sensors 2022;24: 100565.
[32] Gasparini L, Shepherd DA, Bavin EL, Eadie P, Reilly S, Morgan AT, et al. Using machine-learning methods to identify early-life predictors of 11-year language outcome. Journal of Child Psychology and Psychiatry 2022;64:1242−52.
[33] Murugan K, Cherukuri NK, Donthu SS. Efficient recognition and classification of stuttered word from speech signal using deep learning technique. In: 2022 IEEE world conference on applied intelligence and computing (AIC). IEEE; June 2022. p. 774−81.
[34] Arpitha V, Samvrudhi K, Manjula G, Sowmya J, Thanushree GB. Diagnosis of disordered speech using automatic speech recognition. International Journal of Engineering Research and Technology 2020:126−32.

[35] Gkeka E, Agorastou E, Drigas A. Artificial techniques for language disorders. International Journal of Recent Contributions from Engineering Science and IT 2019;7(4): 68–76.

[36] Sheikh SA, Sahidullah M, Hirsch F, Ouni S. Machine learning for stuttering identification: review, challenges and future directions. Neurocomputing 2022:385–402.

[37] Anthony AA, Patil CM, Basavaiah J. A review on speech disorders and processing of disordered speech. Wireless Personal Communications 2022:1–11.

[38] Hegde S, Shetty S, Rai S, Dodderi T. A survey on machine learning approaches for automatic detection of voice disorders. Journal of Voice 2019;33(6). 947-e11.

[39] Yao D, Chi W, Khishe M. Parkinson's disease and cleft lip and palate of pathological speech diagnosis using deep convolutional neural networks evolved by IPWOA. Applied Acoustics 2022;199:109003.

[40] Mathad VC, Liss JM, Chapman K, Scherer N, Berisha V. Consonant-vowel transition models based on deep learning for objective evaluation of articulation. arXiv 2022. preprint arXiv:2203.10054.

[41] Huqh MZU, Abdullah JY, Wong LS, Jamayet NB, Alam MK, Rashid QF, et al. Clinical applications of artificial intelligence and machine learning in children with cleft lip and palate—a systematic review. International Journal of Environmental Research and Public Health 2022;19(17):10860. 1-17.

[42] Evangeline B. A survey on artificial intelligent based solutions using augmentative and alternative communication for speech disabled. 2022. p. 1–26. https://doi.org/10.21203/rs.3.rs-2225081/v1.

CHAPTER

8 Machine learning for cognitive treatment planning in patients with neurodisorder and trauma injuries

Ramandeep Sandhu[1], Deepika Ghai[2], Suman Lata Tripathi[2], Ranjit Kaur[3], Kirti Rawal[2] and Kanav Dhir[4]

[1]*School of Computer Science & Engineering, Lovely Professional University, Phagwara, Punjab, India;* [2]*School of Electronics & Electrical Engineering, Lovely Professional University, Phagwara, Punjab, India;* [3]*School of Computer Applications, Lovely Professional University, Phagwara, Punjab, India;* [4]*Department of Chemistry, S.C.D Government College, Ludhiana, Punjab, India*

1. Introduction

Trauma injuries may contribute to a patient's neurologic condition. It is a type of damage to the nervous system of a person that may also have effects on the rest of the body. It might lead to a person being physically sold off or it might be to blame for a person's motion system being damaged [1–3]. Diseases associated with the peripheral as well as central nervous systems are identified as neurological disorders. Brain disorders or conditions affecting the system that connects the human body's nerves, spine, and brain are related to neuro illnesses. Some problems carry a high risk, some are manageable, and some fall into the category of not causing long-term harm to the body or nervous system [4,5].

Injuries to the brain, multiple sclerosis, Parkinson's disease, epilepsy, dementia, headache disorders, Alzheimer's disease, and neuroinfections are just a few of the more than 600 illnesses that have an impact on the nervous system. The entire nervous system can be affected by a number of bacterial, fungal, viral, and parasite illnesses, including Neisseria meningitides, Zika, West Nile virus, and enteroviruses, as well as fungal and fungus-related infections including Aspergillus and Cryptococcus. The aforementioned neurological symptoms could be caused by an illness itself or an immunological reaction to it. Neurological problems affect hundreds of millions of individuals globally [6–8]. Fig. 8.1 depicts the various kinds of neurological diseases.

In today's high-thinking end world, neurological disorders in humans are becoming more and more prevalent. As the spread affects practically everyone, including adults, children, and healthy infants, it is a severe problem [9,10].

Computational Intelligence and Deep Learning Methods for Neuro-rehabilitation Applications
https://doi.org/10.1016/B978-0-443-13772-3.00012-1

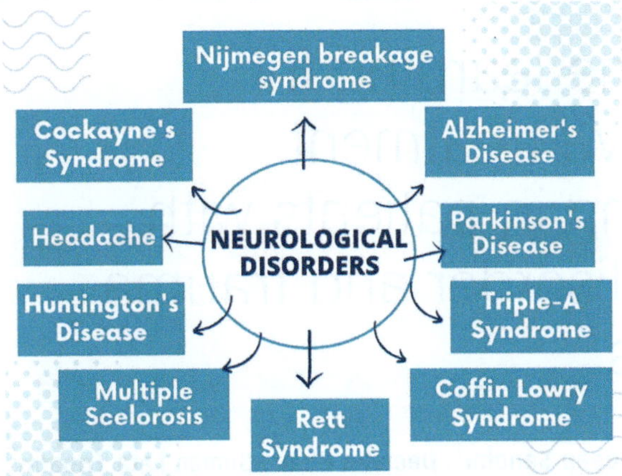

FIGURE 8.1

Common neurological disorders.

A few of the indications and symptoms of neurological disordered people include variable degrees of paralysis, the absence of cutaneous sensation, muscle twitching, convulsions, discomfort, clumsiness, and no presence of skin sensation [11,12]. Fig. 8.2 depicts the signs and symptoms of neurological diseases.

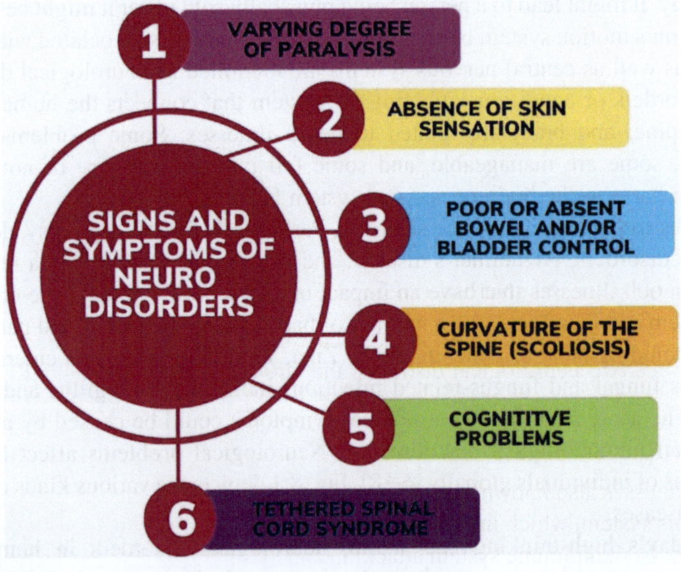

FIGURE 8.2

Signs and symptoms of neurodisorders.

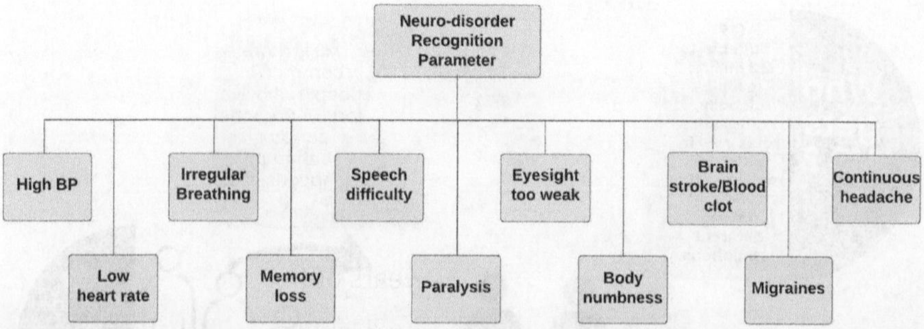

FIGURE 8.3

Neurodisorder recognition parameters.

The neurodisorder recognition parameters such as high BP, low heart rate, and body numbness as shown in Fig. 8.3.

An aerobic and weight-bearing exercise program, in addition to medications, can boost energy levels, lessen apathy, and enhance overall health. To assure participation in a fitness program that involves safe standing and walking once again, a personal trainer or carer may be useful [1–5]. Various neural disorders are enlisted as follows.

- Epilepsy
- Multiple sclerosis (MS)
- Postherpetic neuralgia
- Brain tumors
- Memory disorders
- Headache
- Arteriovenous malformation
- Stroke
- Peripheral neuropathy
- Parkinson's disease (PD)
- Brain aneurysm
- Spinal cord tumor
- Dural arteriovenous fistulae
- Amyotrophic lateral sclerosis (ALS)

Multiple sclerosis and Parkinson's disease, which impact the majority of sufferers, are the two neurological disorders that are explained below.

1.1 Multiple sclerosis

Chronic illness is also known as multiple sclerosis. It has a harmful effect on the central nervous system, which includes the brain, spinal cord, and optic nerves. According to experts, the immune system attacking and harming the nerve system results in MS, an autoimmune disease [13,14]. With alternating intervals of diminution, good health, and incapacitating flare-ups, it often advances gradually. The majority of

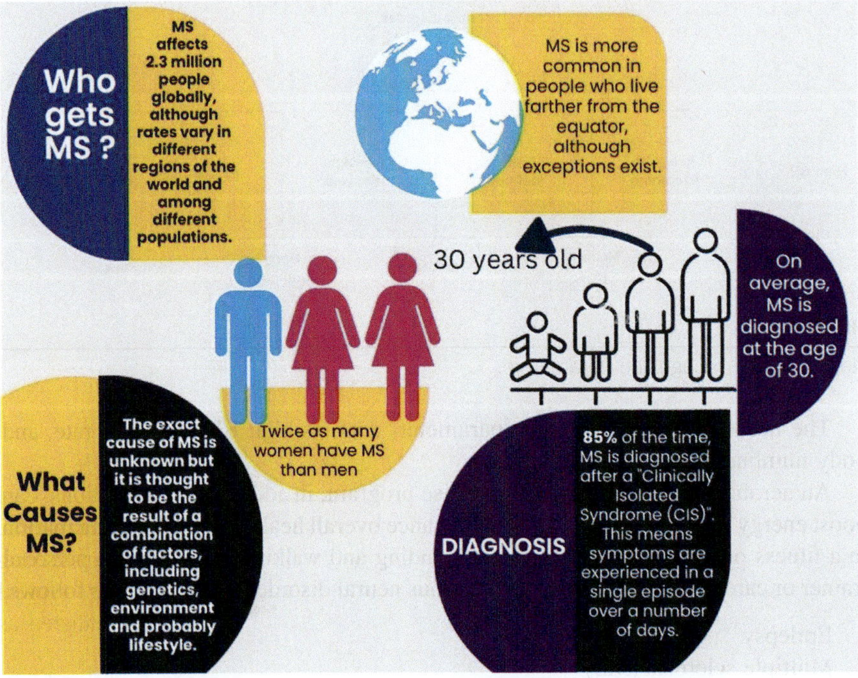

FIGURE 8.4

Causes and diagnosis of multiple sclerosis.

400,000 Americans with MS are young individuals. Women experience it almost two times more often than men. Although there is no known, long-term cure for MS, the majority of people live normal lives and are active for many years following their diagnosis. But, MS can be exhausting, necessitating changes to one's routine and way of life. Some symptoms can be lessened with a daily routine-based exercise regimen that comprises walking, swimming, stretching, and/or riding a motionless bike. By enhancing the MS patient's walking ability, range of motion, balance, and stamina, practitioners and carers can teach the patient how to better manage the condition. Fig. 8.4 shows the causes and diagnosis of multiple sclerosis.

1.1.1 Multiple sclerosis signs and symptoms

Depending on whatever specific nerve fiber pathway is destroyed, different MS symptoms will manifest. When myelin in the spinal cord is destroyed, symptoms including tingling, numbness, tightness, or weakness may appear. Urinary incontinence may happen if the bladder's nerve fibers are compromised [13,14]. Similar to how injury to the brain's cerebellar region can cause an imbalance or a loss of coordination. Depending on where the central nervous system is damaged, MS sufferers might have a wide range of symptoms. Symptoms typically manifest in adults between the ages of 20 and 40.

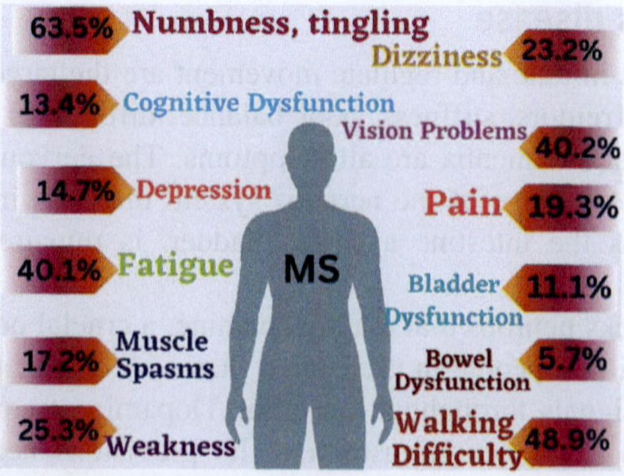

FIGURE 8.5
Signs of multiple sclerosis.

Doctors frequently have trouble diagnosing MS since it can cause a wide range of symptoms in different patients. Changes in muscular function or motor symptoms, such as difficulties walking, stiffness, or tremors, as well as changes in feeling or sensory symptoms like tingling and numbness, are however typical signs of MS. Some common signs of multiple sclerosis (Fig. 8.5) are enlisted as follows.

- Difficulty walking or maintaining balance
- Tingling or numbness
- Eye complications such as uncontrolled eye movements or sometimes double vision
- Depression or emotional charges
- Muscle spasms
- Sexual problems
- Dizziness or vertigo
- Fatigue
- Clumsiness or weakness
- Slight intellectual change such as memory problems
- Stiffness
- Tremors
- Problems with bladder or bowel control

1.1.2 Treatment of multiple sclerosis

Typically, a treatment plan is created by a neurology team in conjunction with the patient's primary care physician, physical therapist, and other healthcare providers. Depression, exhaustion, soreness, stiffness, and tremors are just a few of the symptoms that can be treated or controlled with medication. Physical therapy and exercise, occupational therapy to support independence in everyday life, and stress reduction are all possible treatment programs and services.

1.2 Parkinson's disease

The brain cells that initiate and regulate movement are the targets of PD, a neurological condition. Tremors, stiffness, poor balance, difficulties speaking, sadness, cognitive decline, and dementia are all symptoms. The nervous system, which is "autonomic," the percentage of the nervous system that aids in controlling blood pressure as well as the intestine and the bladder, is impaired in some people [15–17].

The illness attacks neurons that make dopamine, a crucial neurotransmitter that affects around 1 million people in the United States. A chemical messenger called dopamine carries signals throughout the brain. Dopamine frequently functions in mild stability with other neurotransmitters. Its provisions help synchronize the lots of nerve and muscle cells engaged in the movement. Parkinson's sufferers' loss of dopamine-producing brain cells affects how well their nerve and muscle cells coordinate [18,19].

1.2.1 Signs and symptoms of Parkinson's disease

Parkinson's disease typically develops slowly, and symptoms may not show up for years. Although it has been identified as early as age 20, the disease typically affects persons over the age of 50. A familial history of PD runs in about 15% of PD patients. Most people live productive lives for many years after receiving a diagnosis because the disease progresses gradually [20,21].

Following are a few of the early signs of Parkinson's disease that are frequently felt.

- Legs and arms stiffen and become challenging to move.
- Excessive trembling in the legs, arms, or hands is known as a tremor.
- Movements that are slowed down and difficult to begin or complete are referred to as bradykinesia.
- Impaired balance is the inability to maintain equilibrium or to quickly adapt to changes in position.

The PD patient may find it challenging to pick up, handle objects, eat, walk, write, or react faster to avoid injury if they fall due to these symptoms.

Erectile dysfunction, trouble sleeping, problems peeing, problems speaking or swallowing, drooling, hunched posture, an inability to make facial expressions, oily skin, constricted handwriting, constipation, excessive sweating, shortness of breath, and anxiety are other signs and symptoms. Additionally, they are frequently quite difficult to understand because their speech tends to be subdued and rambling. Communication with others is impacted by this. Parkinson's patients frequently struggle to begin, whether speaking or walking. Mobility and movement are quite challenging for people with PD [15–17,20,21].

1.2.2 Treatment process of Parkinson's disease

Parkinson's symptoms can be significantly reduced with medication, but there is currently no cure for the illness. Over time, drugs typically become less effective.

Surgery is a possibility in some situations. The majority of medical professionals advise physical treatment and muscle building. The best results are often achieved for patients by combining walking with active weight-bearing exercise, medication management, and physical therapy.

1.2.3 Exercise and alternative treatments

The finest of conventional medicine and complementary therapies, such as meditation, acupuncture, massage, herbs, and nutrition, are combined in complementary or integrative care. Numerous therapies, including straightforward exercise, physical therapy, massage, and support groups, can reduce symptoms and enhance the quality of life.

Exercise appears to be helpful for patients with Parkinson's disease, according to a recent study. You may ask not only what kind of exercise is best, but also how severe the activity should be. There is mounting evidence that rigorous exercise may help persons with Parkinson's disease by protecting their brains.

Additionally, older women who engage in long-term, regular physical activity, such as walking, have a much higher cognitive function and less cognitive deterioration.

1.3 Importance of artificial intelligence in healthcare

The health society has adopted technical advancements such as ML, AI, IoT, and cloud computing for a safe future health system. Artificial intelligence's subfield, machine learning, is highly recommended as it is a base pillar for a secure healthcare sector. Here, the ML has further supported like smart health records (i.e., electronic health record [EHR]), betterments in clinical tests, magnetic resonance imaging (MRI) image processing techniques, patient data collection and maintenance, and prediction about patient health. Machine learning techniques are highly effective in the treatment of various kinds of chronic diseases such as arthritis, cancer, stroke, and asthma [10,11].

There are various algorithms available in a technological society. If we take the example of the artificial intelligence (AI) field, then it is one of the types of algorithms for a better future with a better healthcare system as it provides various medical facilities as shown in Fig. 8.6. We have considered the brain disorders-related health issues such as brain stroke, which also has a harmful impact on the full body of the patient.

Classifying diseases, especially relevant to the brain system, is one of the most staid and alive or death-based concepts. Various kinds of resourceful research have been already performed. These researchers have been done through high-tech collaborations among researchers in the medical field and technological fields. Still, it is a major challenge to crack brain-related issues of persons. It has been studied that more chances of occurrence of brain stroke (a neurodisorder problem) happens at the age of 30. Exceedingly talented techniques are used by neuroexpert doctors for providing the best treatment based on highly relevant

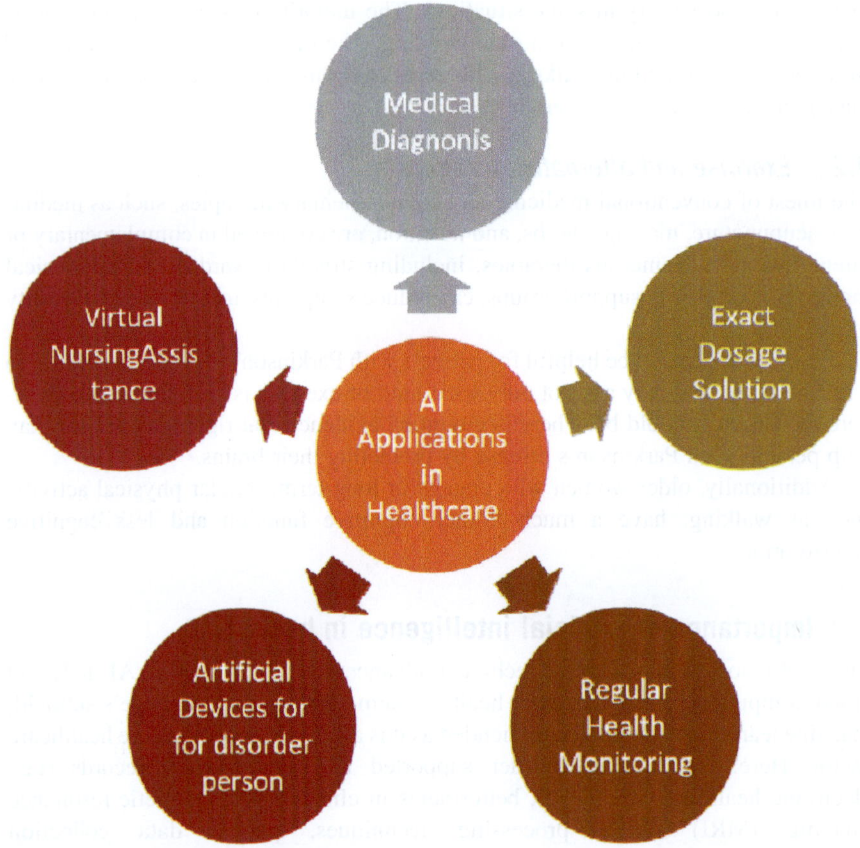

FIGURE 8.6

Various AI applications in healthcare.

scanned diseases. Now if we take up the concern of neuro people or those who are living their life but having some disorder in their neuro phase like brain stroke, some past injury in the brain, having some trauma, etc., these are the factors that arise with some disordering problem of the brain but can harm any body part like paralyze (most probably one side of the body), swelled tongue and so facing as speechless (not a permanent disease but kind of attack on tongue movement), swelling (most chances are mouth swelling), hearing problem, etc. In this chapter, the Kaggle dataset has been used for stroke prediction with the assumption that various impactful parameters have some relevance with neurodisorder. Fig. 8.7 represents the basic components of a human brain system (HBS), which is further divided into two categories.

According to HBS representation as depicted in Fig. 8.7, central nervous system (CNS) and peripheral nervous system (PNS) are two categories of the HBS.

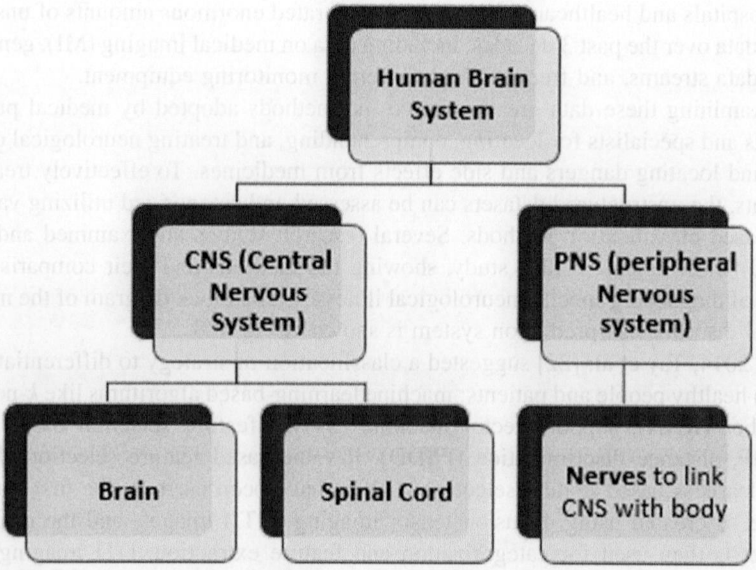

FIGURE 8.7
Human brain system.

Brain functioning and the spinal cord are linked with the CNS system. All nerves link the brain with the body system of a human. Now, if a patient suffers from a brain stroke, then chances are that it may harm the body system also.

2. Related work

In this era of technology, AI, ML, DL, IoT, and cloud computing have taken a great step toward the final relevant and best techniques for patients who are suffering from neurological disorders. The medical community currently faces significant challenges when trying to diagnose various neurodisorders such as epilepsy, Parkison's disease, Alzheimer's disease, multiple sclerosis, acquired brain injury, and ischemic brain stroke in its early stages. If a person receives an early diagnosis, they can improve their quality of life even if their health deteriorates. It is challenging since many neurological diseases have symptoms with other conditions, such as fatigue, tremors, seizures, discomfort, loss of coordination, and consciousness making this possible for a neurological ailment to go unnoticed or, worse, to be incorrectly diagnosed.

According to related research, radiologists, neurosurgeons, and other medical professionals can benefit from automated computer-aided diagnosis (CAD) systems based on AI and ML approaches for improved clinical-related decision-making and the early diagnosis of neurological disorders.

Hospitals and healthcare systems have generated enormous amounts of unstructured data over the past 3 decades, including data on medical imaging (MI), genomic data, data streams, and free text from different monitoring equipment.

Examining these data greatly altered the methods adopted by medical professionals and specialists for locating, comprehending, and treating neurological disorders and locating dangers and side effects from medicines. To effectively treat the patients, the unstructured datasets can be assessed and categorized utilizing various ML-based classification methods. Several research studies are examined and presented below as part of this study, showing the methods and their comparison in terms of diagnosing specific neurological illnesses. The block diagram of the neurological disorder risk prediction system is shown in Fig. 8.8.

In 2014, Tay et al. [22] suggested a classification of strategy to differentiate between healthy people and patients; machine learning-based algorithms like *k*-nearest neighbor (KNN), support vector machine (SVM), feature selection based on a ReliefF, distance discrimination (FSDD), R-value-based feature selection (RFS), and clearness-based feature selection (CBFS) have been used. In the first stage, a dataset is created using diffusion tensor imaging (DTI) images, and the resulting dataset is then used for categorization and feature extraction. DTI imaging data were obtained from the Dankook University Hospital to create the training dataset

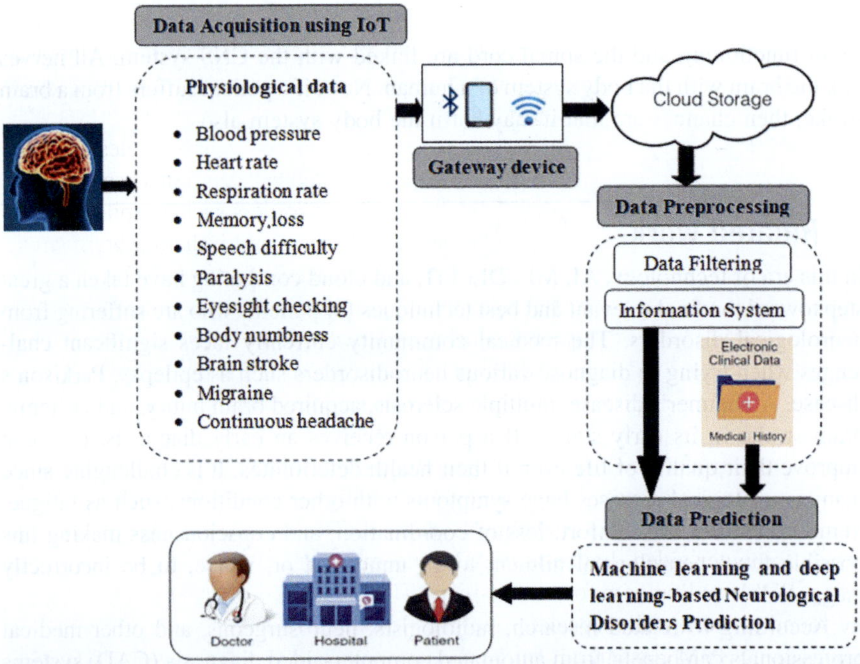

FIGURE 8.8

Block diagram of neurological disorder risk prediction system.

(Cheonan, Republic of Korea). The test data were used to do prediction analysis after creating the training dataset. After that, Te SCI was clinically assessed using ISNCSCI standards. Matlab Version 2007a and MedINRIA Version 1.8.0 are used to extract fractional anisotropy (FA) values from the digital imaging and communication in medicine (DICOM) image. The suggested research contributes to the diagnosis of spinal cord injuries in the C4–C6 region.

In 2017, Serra et al. [23] investigated machine-learning approaches such as probabilistic learning, instance-based learning, decision tree learning, *k*-nearest neighbor classifier, Naïve Bayes classifier, classification-regression tree classifier (CART), SVM named radial basis function kernel (SVM_R), polynomial kernel of degree 2 (SVM_P), and linear kernel (SVM_L) for the automatic cognitive prognosis of patients having a brain injury. It is stated that the improvement of patients in any one of three cognitive domains-attention, memory, or executive functioning is predicted using a variety of classifiers. The research demonstrates that pretreatment diagnosis information can predict how well acquired brain injury (ABI) patients would respond to cognitive training. It is also mentioned that the age of the patient at the time of damage, the patient's underlying cause, or the results of a neuropsychological examination that are gathered before treatment are all proven useful for prognosis and readily produce statistically significant accuracy. The implementations used by researchers are taken from the scikit-learn package (version 0.10) and its default parameters are used. On the distinct accuracy values of count 200 obtained for each fold, researchers utilized the Wilcoxon signed rank: a nonparametric-based statistical hypothesis test to establish statistical significance. When comparing the population mean ranks of two related samples (or two repeated measurements on a single sample), the Wilcoxon signed-rank test is utilized.

In 2019, Raghavendra et al. [24] explored and reviewed recent articles on artificial intelligence and machine learning-based automated CAD systems used over the last 2 decades for the detection of five neurological disorders such as epilepsy, Alzheimer's disease, multiple sclerosis, Parkison's disease, and ischemic brain stroke using physiological signals and images, reviewed in detail. The use of AI approaches for automated seizure detection and epilepsy diagnosis was summarized in this work. The SVM and back propagation neural network are regarded as being the more widely used classifiers. Additionally summarized is the study on automated PD diagnosis utilizing several modalities. It was discovered that the discrete wavelet transform (DWT) and higher order spectra (HOS) combined with entropy and energy features performed effectively. Discrete wavelet transform (DWT) and various dimensionality reduction approaches, including summarization, have been employed extensively for Alzheimer's disease. The majority of ischemic brain stroke computer-aided diagnosis (CAD) systems were tested utilizing fuzzy clustering, SVM, and random forest classifiers. The generally used classifier techniques are found to work well, especially the gray-level co-occurrence matrix-based techniques using SVM classifier. It has been suggested by researchers that CAD systems that have been trained using a significant amount of patient data, physiological signals, and images, based on a skillful combination of advanced signal processing and

AI/ML strategies in an automated manner, potentially help neurosurgeons, neurologists, radiologists, and other healthcare professionals in making better clinical decisions.

In 2020, Segato et al. [25] proposed a detailed review of various artificial intelligence techniques being used in the brain care area. Scopus, Pubmed, and WoS databases are explored in depth to gather information for both the quantitative and qualitative analysis of major artificial intelligence-based algorithms primarily used for diagnosis, surgical treatment, intraoperative assistance, and postoperative assessment. SVM and random forest (RF) machine learning approaches are also studied. According to the authors, artificial intelligence-based analytical tools can assist clinicians in the effective and efficient analysis of the vast quantity of unstructured data.

In 2022, Lima et al. [26] formulated a crisp and detailed study of various challenges, which occur during the disorder detection phase along with the review of classification algorithms. This research article can be taken as a base by the researchers who are interested in the study of neurodiseases. Computer-aided diagnosis methodology, which leads to preprocessing and feature extraction techniques, is also exhibited. A critical review and comparison of the existing ML and DL approaches for the diagnosis of neurological disorders, modalities, and disease-specified datasets that detect and reorder images, signals, speeches, etc. are depicted in the article. This article also provides an outline of the standard evaluation metrics for efficient and effective result analysis. Present methods for developing different neurological disorders systems are also summarized in the article.

Fusco et al. [27], since deferments in patient discharge, can severely impact the efficiency of hospitals, and cost parameters of healthcare can be increased. Therefore, this paper focuses on the early predictors of outcome for discharge, especially in older grown-ups with severely developed brain injuries. A total of 22 patients who are hospitalized in a rigorous neurorehabilitation facility between June 2019 and December 2021 are the subject of a retrospective study. The declaration of Helsinki and the international strategies for better clinical practice are followed during the conduct of this investigation. Based on discharge destination, patients are categorized into two outcome groups: good outcome (GO) or poor outcome (PO), and the potential prognostic factors are examined 1 and 2 months after admission.

Each patient's cognitive and behavioral status is evaluated using the following outcome measures: level of cognitive functioning (LCF), disability rating scale (DRS), and modified Barthel index (MBI), which evaluates independence in daily living activities. The following factors are specifically identified, that is, gender, age, general clinical conditions, time since the injury, severe acquired brain injury etiology, lowest GCS recorded in acute care division, presence of disorder of consciousness (DOC), and body mass index (BMI), including the unresponsive wakefulness syndrome (UWS) and minimally conscious state (MCS). As per the study, an examination of dynamic predictors may be useful to ensure an early evaluation of hospital release in vulnerable individuals with severe acquired brain injury. Additionally, the potential occurrences of consequences are noted, including sepsis, critical illness myopathy and neuropathy (CRIMYNE) syndrome, and pressure sores. Stata software was used to perform statistical analysis.

Rana et al. [28] analyzed a literature review to give a detailed study of the use of a total of 217 research publications published up until September 2022 to analyze a variety of datasets, methodologies, features, modalities, experimental arrangements, and architectural designs that have been used in the illness diagnosis of the subjective ailment. According to the article, it is observed that to solve and enhance the PD diagnostic process, methods such as AI, ML, and DL proved effective and efficient for the classification of normal controls (NC) and PD with comparable therapeutic presentations. It has also been found that AI and ML-based techniques assist physicians in the medical decision-making process, which can lead to the in-depth detection of Parkinson's disease. Last, but not least, for upcoming Parkinson's research, the authors discussed the challenges and offered suitable recommendations for selecting strategies that could be used for connection analysis and subgrouping of data from structural magnetic resonance imaging (SMRI), daTSCAN, and single-photon emission computerized tomography (SPECT). This study's primary goal is to synthesize and evaluate the analysis of artificial intelligence algorithms, data collection techniques, and applications in the diagnosis of arbitrary illnesses and problems.

Fonseca et al. [29] examined the propensity to predict mortality after traumatic brain injury (TBI) of four machine learning techniques, namely extreme gradient boosting (XGBoost), random forest (RF), KNN, and artificial neural networks (ANN). It is concluded that feature selection can enhance the quality of prediction by pairing external feature selection techniques with models like KNN or with decision-tree-based models with built-in feature selection capabilities, like XGBoost, by comparing diverse combinations of machine learning models and feature selection tools. Additionally, feature selection increases the approaches' understandability and makes it easier to compare their clinical relevance and predictive performance. Despite the experimental nature of this research, the results showed that machine learning approaches could profit from the data in intensive care unit (ICU) records, enabling the extremely accurate prediction of death in pediatric TBI patients.

From the literature review, it is concluded that a great variety of current clinical issues relating to the brain are increasingly finding effective solutions through AI applications. Particularly, it has been feasible to considerably improve our understanding of complicated brain mechanisms in recent years, mainly to the accumulation of pertinent data and the development of ever-more-efficient algorithms. The researchers' work is producing more complex and understandable algorithms, which may promote greater usage of "intelligent" technologies in real-world healthcare settings. New opportunities for predicting and managing numerous neurological illnesses have emerged to developments in high-speed computing techniques and impressive advancements in developing fresh DL-based methodologies and models. After reviewing the related research, it is discovered that while many AI, ML, and DL techniques are frequently used in the medical field to accurately diagnose neurological disorders at an early stage, there are still many areas that have not yet been extensively researched but have a significant amount of potential. The comparative analysis of the existing state-of-the-art methods on neurological disorders is shown in Table 8.1.

Table 8.1 Comparative analysis of the existing state-of-the-art methods.

Author (Year)	Techniques, datasets, remarks
Tay et al. [22] (2014)	Techniques: • KNN and SVM • Feature selection based on distance discrimination (FSDD) • R-value-based feature selection (RFS) • ReliefF • Diffusion tensor imaging (DTI) • clearness-based feature selection (CBFS) Datasets: NA Remarks: • DTI technology has been used in the investigation of spinal cord injury (SCI). • Four fractional anisotropy (FA) values from DTI slice pictures of patients and healthy controls analyzed and a technique is described for creating datasets for diagnosis. • The extended dataset enhanced prediction accuracy through feature selection. • Due to legal restrictions on collecting medical data, it is a weakness of the study that researchers were unable to collect a significant amount of SC data. • The C4–C6 region-specific dataset is intended for certain SCI instances. • A new kind of dataset will be needed for a particular type of injury. The study's ultimate objective is to use a predefined training dataset to predict damaged SC segments.
Serra et al. [23] (2017)	Techniques: • Decision-tree learning • Instance-based learning • Probabilistic learning • Naive Bayes • KNN • Classification and regression tree classifier (CART) • SVM: linear kernel (SVM_L), radial basis function kernel (SVM_R), and polynomial kernel of degree 2 (SVM_P). Datasets: • Diagnosis data: From PREVIRNEC, a web-based platform for telerehabilitation designed to support cognitive rehabilitation. ➢ A group of 503 patients is taken into consideration; all were tested for executive functions (501), memory (317 improved), and attention (299 improved) (334 improved).

Table 8.1 Comparative analysis of the existing state-of-the-art methods.—*cont'd*

Author (Year)	Techniques, datasets, remarks
	• Clinical data: 17 neurophysiological tests. • Demographic data: Gender and age of the patient at the time of injury Treatment data: age treatment, delay treatment, treatment weeks, and sessions per week. Remarks: • Improvement of patients in any one of three cognitive domains: attention, memory, or executive functioning is predicted using a variety of classifiers. • Pretreatment diagnosis information can predict how well ABI patients would respond to cognitive training. • It is observed that gender never reaches a single level of statistical significance. • Therefore, it can be viewed as irrelevant for prognosis when taken alone. • Findings show that age at injury and age treatment are two highly predictive notions with statistically significant accuracy levels typically exceeding 55%. • More recoveries than nonrecoveries in patients under the age of 37, and more nonrecoveries than recoveries in patients over the age of 57. • To allow for a finer refinement, using days or weeks as units rather than years is recommended. • There is not much of an increase in accuracy when general and specific etiologies or general and specific preevaluation scores are combined. • It is observed that a larger number of classifiers acquire statistically meaningful accuracies when etiologies or preevaluations are combined. • Finally, a quick examination of feature relevance was conducted. • The selected feature relevance analysis methods were executed on the pool of all available features and display the top 10 features for each one. • Researchers concentrated on automatic cognitive prognosis, or determining whether a patient will progress in a variety of cognitive abilities using only pretreatment data. • Research demonstrates the relevance of factors including injury age, etiology, and neuropsychological exam results for prognosis, producing statistically significant prediction accuracies. • The fact that the results are independent of the classification scheme emphasizes the predictive ability of the variables under consideration. • Additionally, preevaluation tests and the researched cognitive functions appear to have hidden and emergent relationships that can be discovered using machine learning approaches.

Continued

Table 8.1 Comparative analysis of the existing state-of-the-art methods.—*cont'd*

Author (Year)	Techniques, datasets, remarks
	• The authors intend to add therapy data to our analysis in upcoming work. • By fusing therapy and diagnosis data in this manner, new patient outcomes from a pool of past patients can be improved.
Raghavendra et al. [24] (2019)	Techniques: • ML-based automated CAD system • SVM • Back propagation NN • DWT • HOS • Fuzzy clustering • Random forest (RF) • Deep learning techniques: supervised and unsupervised • K-means clustering Datasets: NA Remarks: • Artificial intelligence and machine learning based automated CAD systems used over the last 2 decades for the detection of five neurological disorders such as epilepsy, Alzheimer's disease, Parkison's disease, ischemic brain stroke using physiological signals and images, and multiple sclerosis reviewed in detail. • Epilepsy is the subject of automated CAD system research more often than PD, and AD. • They rely on EEG signals for seizure detection and epilepsy diagnosis. • They are created for PD utilizing a variety of modalities, including EEG, voice, handwriting, and brain imaging. • Artificial intelligence (AI)/Machine learning (ML) techniques in an automated manner can help neurologists, neurosurgeons, radiologists, and other healthcare professionals to make better clinical decisions. • DWT and various dimensionality reduction techniques have been widely applied to the study of Alzheimer's disease. • Fuzzy clustering, SVM, and random forest classifiers are used in the majority of ischemic brain stroke computer-aided diagnosis (CAD) systems. According to the authors of the article, one of three directions is possible for future research: The development of a sizable public database for the validation of all developed techniques, early neurological disorder detection using ML-based CAD, and remote access for clinicians without expertise.
Segato et al. [25] (2020)	Techniques: • Support vector machine • Random forest • Diffusion tensor imaging • Fluid attenuated inversion recovery (FLAIR) • Functional magnetic resonance imaging (fMRI) • Hyperspectral imaging

Table 8.1 Comparative analysis of the existing state-of-the-art methods.—*cont'd*

Author (Year)	Techniques, datasets, remarks
	• Linear regression algorithm • Logistic regression algorithm • Instance-based algorithms • k-nearest neighbor • Bayesian algorithms Datasets: • Medical-imaging • MRI • MRI-DWI (diffusion weighted imaging) • MRI-FLAIR (fluid-aattenuated inversion recovery) • MRI-DTI (diffusion tensor imaging) • Computed tomography (CT) • Microelectrode recording (MER) • Electroencephalography (EEG) • Positron emission tomography (PET) • Intra-operative ultrasound (IUS) • Electronic health record (EHR) Remarks: • A detailed review of various artificial intelligence techniques being used in the brain care area is presented. • Databases searched to obtain data for the quantitative and qualitative examination of significant AI-based algorithms: Pubmed, Scopus, and Web of Sciences databases. • According to the authors, artificial intelligence-based analytical tools can assist clinicians in the effective and efficient analysis of the vast quantity of unstructured data.
Lima et al. [26] (2022)	• Machine learning and deep learning methods • DWT • DCT • DL-based brain tumor classification (BTC) • Histogram normalization • Spatial normalization • Intensity normalization • Z-score normalization • Numerical normalization (NNM) • High-pass-filtering (HPF) • Spatial filtering (SF) • Wiener filtering (WF) • Temporal filtering (TF) • Stripping • Scaling • Distortion correction (DC) • Motion correction (MC) • Slice timing correction (STC) • Smoothing • Linear discriminant analysis (LDA) • Independent component analysis (ICA) • Principal component analysis (PCA) • Support vector machine (SVM) • K-nearest neighbors (KNN)

Continued

Table 8.1 Comparative analysis of the existing state-of-the-art methods.—*cont'd*

Author (Year)	Techniques, datasets, remarks
	• Gaussian mixture models (GMM) • Random forests (RF) • Artificial neural network (ANN) Datasets: • MRI • EEG data • Ultrasound images • Speech recordings • Voice recording • Handwriting • Magnetoencephalography (MEG) • Positron emission tomography (PET) • fMRI • fNIRS • Computed tomography (CT) • Single-photon emission computed tomography (SPECT) • Hyperspectral brain cancer image database • ADNI dataset • CHB-MIT, Freiburg hospital and EPILEPSIAE dataset • UCI dataset • Parkinson's disease dataset • Brain tumor dataset collected from cerebrum web informational index • Population-based nested case-control study • University of California at Irvine (UCI) machine learning repository • Clinical and HRV data • Parkinson's disease spiral drawings using digitized graphics tablet dataset • Molecular brain Neoplasia data (REMBRANDT) • Vasei hospital in Sabzevar for multiple sclerosis Remarks: • A detailed study of various challenges, which occur during the disorder detection phase along with the review of classification algorithms. • Researchers who are interested in the study of neurological illnesses might use this research article as a starting point. • Additionally demonstrated is computer-aided diagnosis methodology, which results in feature extraction and preprocessing methods. • The paper provides a critical analysis and comparison of current ML and DL methods for the diagnosis of neurological disorders, modalities, and disease-specific datasets that recognize and reorder images, signals, utterances, etc. • The article also provides a summary of current techniques for creating systems to treat various neurological illnesses. • Future research is needed in this area that combines different screening techniques to accurately and swiftly diagnose neurological illnesses. • Researchers should concentrate on developing embedded software that improves the precise identification of certain illnesses.

Table 8.1 Comparative analysis of the existing state-of-the-art methods.—*cont'd*

Author (Year)	Techniques, datasets, remarks
Fusco et al. [27] (2022)	Techniques: Focussed on the early predictors of outcome—poor outcome (PO) or good outcome (GO) for discharge. Datasets: • A total of 22 patients (hospitalized in a serious neurorehabilitation facility) from June 2019 to December 2021 were the subject of a retrospective study. Remarks: • Once the intense rehabilitation procedure is complete, being able to predict how patient pathways and outcomes might develop could make discharge simpler and earlier. • The functional and mental state improvements made during the first month of rehabilitation have an impact on the result in older patients upon discharge. • High BMI and sepsis in the first month of hospitalization may also be indicators of a bad outcome, although additional information and a bigger sample size are required to corroborate this finding. • Stata software was used to perform statistical analysis. • It would be interesting to conduct further study on the association between these early dynamic indicators of the outcome at discharge and a patient's functional outcome in terms of residual disability and functional capability.
Rana et al. [28] (2022)	Techniques: • AI • K-nearest neighbor • ML • SVM • DL • ANN • CNN • Fuzzy KNN • Naïve Bayes • Preferred reporting items for systematic reviews and Meta-analyses (PRISMA) technique • Decision tress • Gradient-boosted trees Datasets: • DaTSCAN • Single-photon emission-computerized tomography (SPECT) • Structural magnetic resonance imaging (SMRI) • Screening questionnaire Remarks: • Data was compiled from 217 articles on the diagnosis and categorization of PD published between 2015 and 2022 (from various sources). • Examined the research publications from the Frontiers Media SA (31), MDPI (81), PLOS (23), IEEE (86), nature (9), Hindawi (5),

Continued

Table 8.1 Comparative analysis of the existing state-of-the-art methods.—*cont'd*

Author (Year)	Techniques, datasets, remarks
	Springer (6), Elsevier and nature Portfolio (3) to carry out this study. • Even though there is no known cure for Parkinson's disease, we can slow and even stop its progression with a precise and prompt diagnosis. • AI is a good option for early-stage PD identification when compared to conventional PD detection techniques. • Researchers outlined the current constraints and difficulties and offered potential solutions (recommendations) for the future that could result in effective AI and ML techniques to address the problems.
Fonseca et al. [29] (2022)	Techniques: • KNN • ANN • Random forest • Extreme gradient boosting Datasets: • Hackathon pediatric traumatic brain injury (HPTBI) dataset • Clinical and demographic data of 300 patients Remarks: • The combination of the model and feature selection techniques used has a significant impact on predictive power; however, generally, ML models performed quite well in terms of mortality prediction for traumatic brain injury (TBI). • The findings of the feature importance analysis show that clinical significance and predictive value are not correlated. • This study outlines several limitations that were found as well as potential future research initiatives. • Healthcare dataset sizes typically suffer from a lack of large collections, which has slowed the development and use of AI solutions in the industry. • One of the drawbacks of the current study is specifically the amount of the dataset that was employed. • The learning model may not have generalized sufficiently to account for the population's variability since the dataset may not have captured its heterogeneity. • On the other hand, the application of the technique was constrained by the tiny size of the dataset. • Since deep learning techniques require enormous amounts of training data, they were not used. • Furthermore, the dataset only includes pediatric patients, a population subset with less research on TBI-related classification. • Last but not least, the dataset lacked continuous data, which could be crucial for predicting mortality and enabling the development of novel classification models that take the time sequence into account.

3. Proposed methodology

The process involved in the proposed method of machine learning for cognitive treatment planning in patients with neurodisorder and trauma injuries is shown in Fig. 8.9. It works for a better future life of stroke patients so that no more chance of restroke can occur. The proposed system is a real-time monitoring system. This system is useful for stroke patients so that they can live a safe life.

(a) The first stage reads the complete real-time medical details of neuro disorder/stroke patients using various sensors (i.e., visual sensor, motion sensor, voice record sensor, etc.) or public datasets available online (i.e., Kaggle). The medical details such as eyesight, pulse rate, blood pressure, heart rate, EEG, ECG, temperature, and motion sensing are taken either through sensors or datasets.

(b) The second stage of the proposed method is toward the doctor's prescription to be followed by the patient and store this prescription in the cloud database server. If the patient is not yet discharged from the hospital, it means that the patient is in the hospital and it is under treatment with physical interaction with

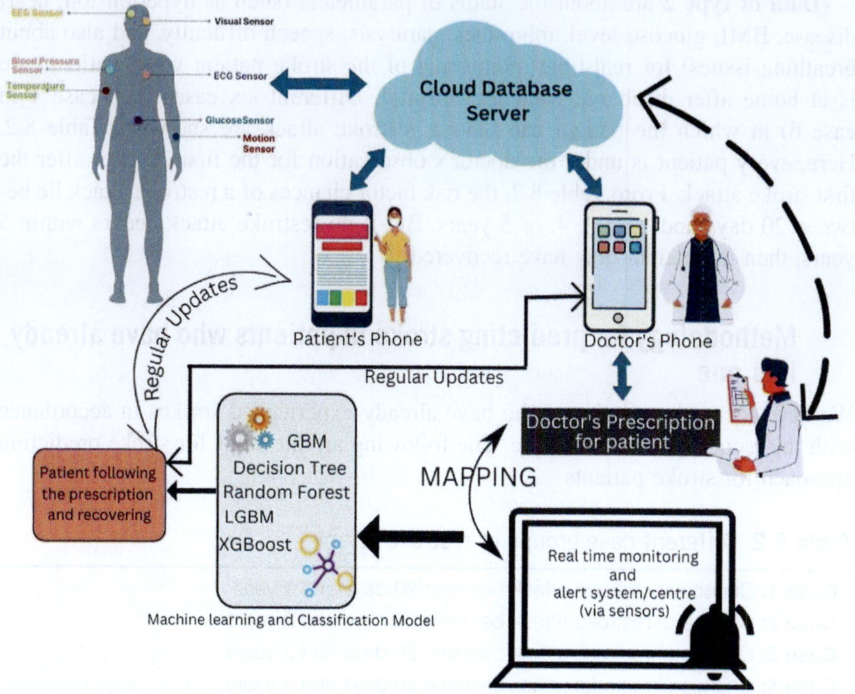

FIGURE 8.9

Diagram of the proposed method of machine learning for cognitive treatment planning in patients with neurodisorder and trauma injuries.

the concerned doctor team. If the patient is discharged from the hospital, then the patient is required to follow the daily routine prescription given by the doctor and the same doctor's prescription information will be available in the cloud database server.

(c) Further, various ML techniques such as decision tree, random forest, KNN, ANN, and SVM are applied for classification based on the mapping in between the real-time patient monitoring through mobile API and the daily routine to be followed by the patient. Now, an alert reminder will be generated as a connection between the concerned doctor and patient using mobile API whether the patient is recovering or not.

To start with the process of better patient care of prestroke patients, two types of primary data of a patient have been considered.

Data of type 1 are static information about patients like age, gender, marital status, working in private, government, or another sector, living in a rural or urban estate, smoking person on not. Apart from this, more parameters related to the patient are carried out to make a setup for better patient care using various sensors such as glucose detection, EEG, and ECG and mobile API such as smart watches and smart bracelets

Data of type 2 are about the status of parameters (such as hypertension, heart disease, BMI, glucose level, migraines, paralysis, speech difficulty, and also about breathing issues) for real-time monitoring of the stroke patient when patient care is at home after discharge from the hospital. Different six cases (i.e., case 1 to case 6) in which the patient can have a restroke attack are shown in Table 8.2. Here, every patient is under the doctor's observation for the first 20 days after the first stroke attack. From Table 8.2, the risk factor chances of a restroke attack lie between 20 days and 1, 2, 3, 4, or 5 years. But if no restroke attack occurs within 5 years, then the patient may have recovered.

3.1 Methodology for predicting stroke in patients who have already had one

We have focused on patients who have already experienced strokes in accordance with the recommended technique. The following are the steps for stroke predicting approach for stroke patients.

Table 8.2 Different case groups of restroke attack.

Case 1: Chance of restroke attack between 20 days and 1 year
Case 2: Chance of restroke attack between 20 days and 2 years
Case 3: Chance of restroke attack between 20 days and 3 years
Case 4: Chance of restroke attack between 20 days and 4 years
Case 5: Chance of restroke attack between 20 days and 5 years
Case 6: If no restroke attack within 5 years, then the patient may be recovered

- Begin the procedure by collecting the information through an EHR, sensor, or publicly available dataset of patients with stroke.
- Verify the patient's discharge status.
  - ➢ If not yet discharged, then accept that the patient is still in the hospital and under proper treatment.
  - ➢ If discharged, then record the patient's discharge date.
- Collect and store doctor's prescriptions on a cloud database server which is provided to the patient at discharge time.
- Mapping of routinely prescribed by a doctor with daily routine followed by the patient based on machine learning algorithms.
- Classification based on supervised machine learning techniques toward patient is recovering or not.
- If the status is recovery, then there is no need to visit the hospital for a check-up and declare the patient is safe from stroke reattack, that is, "No-Dependency-No-Phobia," else visit the neuro hospital for treatment.
- Stop the process.

The flowchart of stroke predictive methodology of patients who have already had one stroke is shown in Fig. 8.10.

4. Results and discussion

We have tested the proposed method on the Kaggle dataset (https://github.com/codejay411/Stroke_prediction/blob/main/Stroke%20Prediction.ipynb), and the patient data are taken for the prediction of whether brain stroke or not. The Kaggle dataset for stroke prediction data is shown in Fig. 8.11. The proposed method runs on a 64-bit operating system—Intel(R) Core(TM) i5-7200U CPU@ 2.50 GHz 2.71 GHz with 8 GB RAM using Python.

The primary parameters of patients, based on the dataset above, are

- id (unique is in integer form)
- patient gender (male, female, or other)
- age (years)
- hypertension (0 or 1- 0 if no, 1 if yes)
- heart disease (0 or 1)
- marital status (either yes or no)
- work-type (private, govt_job, children or self-employed)
- residence type (rural or urban)
- avg_glucose_level (either value or 0)
- BMI (either value or 0)
- smoking status (unknown, formerly smoked, never smoked, smokes)

These parameters have been used to predict whether a patient of this type will experience a stroke or not. According to this Kaggle dataset retrieved from

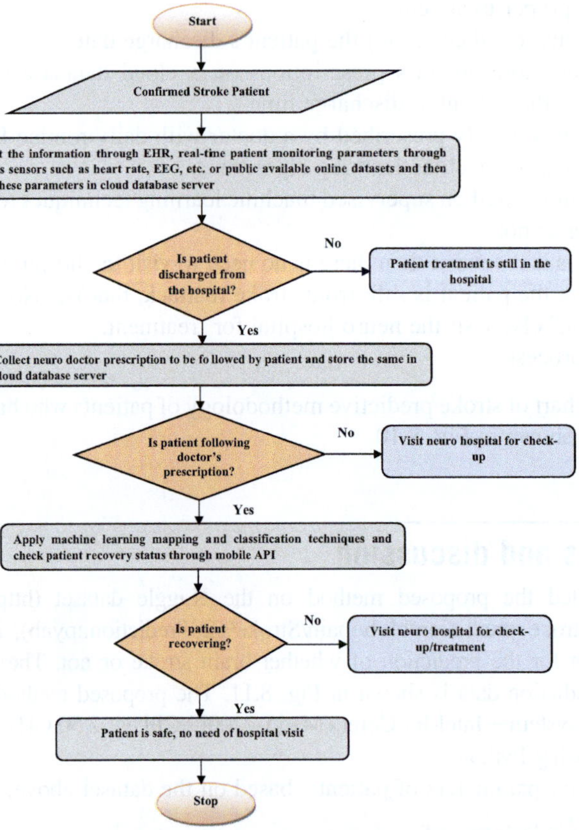

FIGURE 8.10

Flowchart of stroke predictive methodology.

https://github.com/codejay411/Stroke_prediction/blob/main/Stroke%20Prediction.ipynb, it is evident that a survey was conducted using a collection of 4981 rows, each of which contains unique person details and total 11 attributes have been considered. Fig. 8.12 shows the boxplot stroke prediction based on different input attributes such as gender, ever_married, work_type, residence_type, and smoking_status.

id	gender	age	hypertension	heart_disease	ever_married	work_type	Residence_type	avg_glucose_level	bmi	smoking_status
67	Female	17	0	0	No	Private	Urban	92.97	N/A	formerly smoked
77	Female	13	0	0	No	children	Rural	85.81	18.6	Unknown
84	Male	55	0	0	Yes	Private	Urban	89.17	31.5	never smoked
91	Female	42	0	0	No	Private	Urban	98.53	18.5	never smoked
99	Female	31	0	0	No	Private	Urban	108.89	52.3	Unknown
121	Female	38	0	0	Yes	Private	Urban	91.44	N/A	Unknown
129	Female	24	0	0	No	Private	Urban	97.55	26.2	never smoked
132	Female	80	0	0	Yes	Govt_job	Urban	84.86	N/A	Unknown
156	Female	33	0	0	Yes	Private	Rural	86.97	42.2	never smoked
163	Female	20	0	0	No	Private	Rural	94.67	28.8	Unknown
170	Male	43	0	0	Yes	Govt_job	Rural	80.07	N/A	never smoked
172	Male	8	0	0	No	children	Urban	78.76	N/A	Unknown
187	Female	20	0	0	No	Private	Rural	84.07	27.6	smokes
205	Female	43	0	0	Yes	Private	Rural	88.23	37.6	Unknown
210	Male	81	0	0	Yes	Self-employed	Rural	91.54	31.4	never smoked
224	Female	23	0	0	No	Private	Urban	110.16	N/A	never smoked
239	Male	59	1	1	Yes	Private	Rural	246.53	27.2	formerly smoked
242	Male	4	0	0	No	children	Urban	98.56	17.5	Unknown
247	Male	31	0	0	No	Private	Urban	72.6	31.6	never smoked
259	Male	79	0	0	Yes	Private	Urban	198.79	24.9	never smoked
298	Female	41	0	0	Yes	Self-employed	Rural	76.66	N/A	Unknown

FIGURE 8.11

Stroke prediction collected from Kaggle dataset (https://www.kaggle.com/datasets/fedesoriano/stroke-prediction-dataset) of patients who may have chances of stroke occurrence.

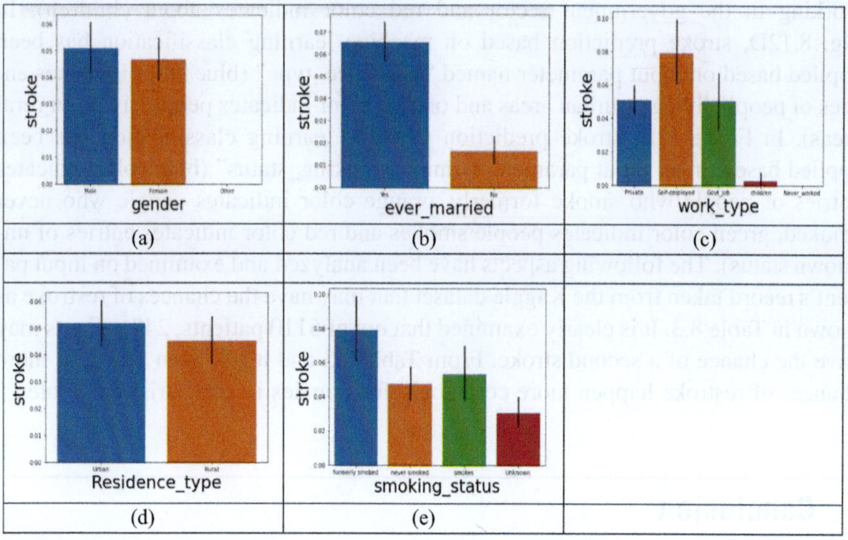

FIGURE 8.12

Stroke chances as per gender, marital status, work type, residence type, and smoking status (https://github.com/codejay411/Stroke_prediction/blob/main/Stroke%20Prediction.ipynb).

Table 8.3 Input patient's record that may have the chances of restroke.

Overall people that may have the chance of restroke	248
Proportion of males	108
Proportion of females	140

On the basis of the input data record from the Kaggle dataset of a total of 5110 people entries (2115 male entries, 2994 female entries, and 1 from the other category), consecutive chances of stroke prediction have been shown in Fig. 8.12. In Fig. 8.12A, stroke prediction has been applied on the basis of input attribute named "gender" (blue color indicates entries of male and orange color indicates female). In Fig. 8.12B, stroke prediction machine learning classification has been applied based on an input parameter named "ever_married" (blue color indicates entries of married people and orange color indicates not-married people). In Fig. 8.12C, stroke prediction machine learning classification has been applied based on input parameter named "work type" (blue color indicates entries of people working in the private sector, orange color indicates self-employed people, green color indicates people working in the government sector and red color indicates about children). In Fig. 8.12D, stroke prediction based on machine learning classification has been applied based on input parameter named "residence_type" (blue color indicates entries of people living in urban areas and orange color indicates people living in rural areas). In Fig. 8.12E, stroke prediction machine learning classification has been applied based on an input parameter named "smoking_status" (blue color indicates entries of people who smoke formerly, orange color indicates people who never smoked, green color indicates people smokes and red color indicates entries of unknown status). The following aspects have been analyzed and examined on input patient's record taken from the Kaggle dataset that may have the chances of restroke as shown in Table 8.3. It is clearly examined that out of 5110 patients, 248 patients may have the chance of a second stroke. From Table 8.3 and it has been seen that more chances of restroke happen more commonly for females in comparison to males.

5. Conclusion

Machine and deep learning models have provided emerging integration into medical treatments. The use of AI, DL, and ML techniques is gradually bringing efficient as well as imaginary solutions to many real-world neurodisorder patients having brain-related issues. In this study, a framework for early detection of stroke reoccurrence with the aid of classification-based machine learning has been provided. Here, the Kaggle dataset of stroke-suffering patients has been considered as a trained model for input. The prediction model is able for early prediction of restroke attacks on the basis of the daily routine followed by the affected patients. The model is able to extract sensor data of stroke patients so that the same information can be stored on the cloud database server. Apart from this, the idea of an optimal real-time monitoring mobile API is suggested. It is connected with both patient and doctor for providing a regular mapping between both of them.

References

[1] D Andrews PJ, Sleeman D, Statham P, McQuatt A, Corruble V, Jones PA, et al. Predicting recovery in patients suffering from traumatic brain injury by using admission variables and physiological data: a comparison between decision tree analysis and logistic regression. Journal of Neurosurgery 2002;97:326−36. https://doi.org/10.3171/jns.2002.97.2.0326.

[2] Hoffman JM, Pagulayan KF, Zawaideh N, Dikmen S, Temkin N, Bell KR. Understanding pain after traumatic brain injury: impact on community participation. American Journal of Physical Medicine & Rehabilitation/Association of Academic Physiatrists 2008;86:962−9. https://doi.org/10.1097/PHM.0b013e31815b5ee5.

[3] Jones JM, Jetten J, Haslam SA, Williams WH. Deciding to disclose: the importance of maintaining social relationships for wellbeing after acquired brain injury. In: Jetten J, Haslam C, Haslam SA, editors. The social cure: identity, health and well-being. New York, USA: Psychology Press; 2012. p. 255−72 [Chapter 14].

[4] Bairy GM, Lih OS, Hagiwara Y, Subha DP, Faust O, Niranjan UCA, et al. Automated diagnosis of depression electroencephalograph signals using linear predictioncoding and higher order spectra features. Journal of Medical Imaging and Health Informatics 2017;7:1857−62. https://doi.org/10.1166/jmihi.2015.1418.

[5] Bhat S S, Acharya UR, Adeli H, Bairy GM, Adeli A. Autism: cause factors, early diagnosis and therapies. Reviews in the Neurosciences 2014;25:841−50. https://doi.org/10.1515/revneuro-2014-0056.

[6] Bhat S, Acharya UR, Adeli H, Bairy GM, Adeli A. Automated diagnosis of autism: in search of a mathematical marker. Reviews in the Neurosciences 2014;25:851−61. https://doi.org/10.1515/revneuro-2014-0036.

[7] Shukla PK, Sandhu JK, Ahirwar A, Ghai D, Maheshwary P, Shukla PK. Multiobjective genetic algorithm and convolutional neural network based COVID-19 Identification in chest X-ray images, vol. 2021; 2021. p. 1−9. https://doi.org/10.1155/2021/7804540.

[8] Ghai D, Gianey HK, Jain A, Uppal RS. Quantum and dual-tree complex wavelet transform-based image watermarking. International Journal of Modern Physics B 2020;34. https://doi.org/10.1142/S0217979220500095.

[9] Ghai D, Tiwari S, Das NN. Bottom-boosting differential evolution based digital image security analysis. Journal of Information Security and Applications 2021;61. https://doi.org/10.1016/j.jisa.2021.102811.

[10] Tripathi SL, Dhir K, Ghai D, Patil S. Health informatics and technological solutions for Coronavirus (COVID-19)". 1st ed. CRC Press; 2021. https://doi.org/10.1201/9781003161066.

[11] Ghai D, Tripathi SL, Saxena S, Chanda M, Alazab M. Machine learning algorithms for signal and image processing". 1st ed. Wiley-IEEE Press; 2022. p. 1−512.

[12] Sridhar C, Bhat S, Acharya UR, Adeli H, Bairy GM. Diagnosis of attention deficit hyperactivity disorder using imaging and signal processing techniques. Computers in Biology and Medicine 2017;88:93–9. https://doi.org/10.1016/j.compbiomed.2017.07.009.

[13] Murray TJ, Bowling AC, Polman C, Thompson A, Noseworthy J. Multiple Sclerosis-the guide to treatment and management". 6th ed. London: Multiple Sclerosis International Federation; 2006.

[14] Kobelt G, Pugliatti M. Cost of multiple sclerosis in Europe. European Journal of Neurology 2005;12:63–7. https://doi.org/10.1111/j.1468-1331.2005.01193.x.

[15] Nutt JG, Wooten GF. Clinical practice. Diagnosis and initial management of Parkinson's disease. New England Journal of Medicine 2005;353:1021–7. https://doi.org/10.1056/NEJMcp043908.

[16] Bhat S S, Acharya UR, Hagiwara Y, Dadmehr N, Adeli H. Parkinson's disease: cause factors, measurable indicators, and early diagnosis. Computers in Biology and Medicine 2018;102:234–41. https://doi.org/10.1016/j.compbiomed.2018.09.008.

[17] Chaudhuri KR, Yates L, Martin PM. The non-motor symptom complex of Parkinson's disease: a comprehensive assessment is essential. Current Neurology and Neuroscience Reports 2005;5:275–83. https://doi.org/10.1007/s11910-005-0072-6.

[18] Marras C, Tanner CM. Epidemiology of Parkinson's disease. In: Watts RL, Kolle WC, editors. Movement disorders, neurologic principles and practice". 2nd ed. New York: McGraw Hill; 2004. p. 177–96.

[19] Shinde S, Prasad S, Saboo Y, Kaushick R, Saini J, Pal PK, et al. Predictive markers for Parkinson's disease using deep neural nets on neuromelanin sensitive MRI. NeuroImage: Clinical 2019;22:1–8. https://doi.org/10.1016/j.nicl.2019.101748.

[20] Zhang A, Segundo RS, Panev S, Tabor G, Stebbins K, Whitford A, et al. Automated tremor detection in Parkinson's disease using accelerometer signals. In: IEEE/ACM international conference on connected health: applications, systems and engineering technologies (CHASE). Washington, DC, USA: IEEE; 2018. p. 13–4. https://doi.org/10.1145/3278576.3278582. Sept. 26–28.

[21] Gálvez G, Recuero M, Canuet L, Pozo FD. Short-term effects of binaural beats on EEG power, functional connectivity, cognition, gait and anxiety in Parkinson's disease. International Journal of Neural Systems 2018;28. https://doi.org/10.1142/S0129065717500551.

[22] Tay B, Hyun JK, Oh S. A machine learning approach for specification of spinal cord injuries using fractional anisotropy values obtained from diffusion tensor images. Computational and Mathematical Methods in Medicine 2014;2014. https://doi.org/10.1155/2014/276589.

[23] Serra J, Arcos JLI, Rudolph AG, Molina AG, Rovira TR, Tormos JM. Cognitive prognosis of acquired brain injury patients using machine learning techniques. In: 5th international conference on advanced cognitive technologies and applications. Valencia, Spain: IARIA; May 27–June 1, 2013. p. 108–13.

[24] Raghavendra U, Acharya UR, Adeli H. Artificial intelligence techniques for automated diagnosis of neurological disorders. European Neurology 2019;vol. 82:41–64. https://doi.org/10.1159/000504292.

[25] Segato A, Marzullo A, Calimeri F, Momi ED. Artificial intelligence for brain diseases: a systematic review. APL Bioengineering 2020;4. https://doi.org/10.1063/5.0011697.

[26] Lima AA, Mridha MF, Das SC, Kabir MM, Islam MR, Watanobe Y. A comprehensive survey on the detection, classification, and challenges of neurological disorders. Biology 2022;11. https://doi.org/10.3390/biology11030469.

[27] Fusco A, Galluccio C, Castelli L, Pazzaglia C, Pastorino R, Marafon DP, et al. Severe acquired brain injury: prognostic factors of discharge outcome in older adults. Brain Sciences 2022;12. https://doi.org/10.3390/brainsci12091232>.

[28] Rana A, Dumka A, Singh R, Panda MK, Priyadarshi N. A computerized analysis with machine learning techniques for the diagnosis of Parkinson's disease: past studies and future perspectives. Diagnostics 2022;12. https://doi.org/10.3390/diagnostics12112708.

[29] Fonseca J, Liu X, Oliveira HP, Pereira T. Learning models for traumatic brain injury mortality prediction on pediatric electronic health records. Frontiers in Neurology 2022;13. https://doi.org/10.3389/fneur.2022.859068.

CHAPTER

Artifacts removal techniques in EEG data for BCI applications: A survey

9

Srinath Akuthota, K. Raj Kumar and Ravichander Janapati
Department of Electronics and Communication Engineering, SR University, Warangal, Telangana, India

1. Introduction

The rapid advancement of Biomedical Engineering, medicine as well as information technology, and its combination has ushered in the noninvasive electroencephalography-based replica of brain−computer interfaces [1,2]. Many research findings on EEG signals have shown that specific signal bands are strongly associated with specific activities, and the disproportionality in these frequencies is used to diagnose certain disorders and diseases [3,4]. Electroencephalography (EEG) signals, also known as brain waves, are crucial in brain−computer interface (BCI) applications because they show the brain's electrical action. BCI systems are using these brain signals to detect modifications in the brain's electrical activity, which are then converted into control signals to carry out particular tasks, like operating a desktop computer mouse or a robotic prosthetic. Therefore, the success of BCI systems and applications depends on their ability to precisely capture and analyze brain waves.

The different brain waves and their activities are majorly mentioned below.

Gamma Waves: These brain waves are best recorded at the parietal and active frontal lobes of the brain, and these occur when the brain state is mainly perception, attention, and memory, intense focus, concentration, and mental alertness. The major frequency band is 40−400 Hz [5].

Beta Waves: These are the brain rhythms within the frequency range of 4−40 Hz and are best recorded at the parietal lobes and frontal lobes. These are primarily the result of mental abilities such as rationalization, decisions, and solving problems, as well as realization [6].

Alpha Waves: These are the brain signals raised in the occipital lobe and frontal lobe areas of the brain within the range of 8−12 Hz predominantly due to wakeful relaxation, such as during meditation, daydreaming, or during resting state [7].

Theta Waves: These signals range from 4 to 8 Hz, and they are typically strongest during light sleep, deep relaxation, and meditation and are best recorded in frontal and temporal regions of the brain [7].

Computational Intelligence and Deep Learning Methods for Neuro-rehabilitation Applications
https://doi.org/10.1016/B978-0-443-13772-3.00004-2

Delta Waves: These delta rhythms range between 0.5 and 4 Hz, are recognized when the brain state is in deep meditative and trance-like states and deep sleep, and are best recorded in the frontal, central, and parietal regions of the brain [8].

A brain–computer interface, also known as a brain–machine interface, or HCI [9], is the combination of hardware and software for functional connectivity between a biological object and a computer [7], The main interdisciplinary areas of BCI research are neuroscience, signal processing, and clinical research, all of which have connections to artificial intelligence and machine learning techniques [10].

Identification and elimination of nonneural signals could obstruct the part of the BCI artifact separation process; the EEG signals from the brain are analyzed. Muscle activity, eye gaze, and other sources of intervention that are unrelated to brain activity can all be considered artifacts.

The BCI system's operation necessitates the use of three modules (Fig. 9.1).

1. Signal capturing
2. Processing of signals
3. Application interface and applications

1.1 Signal capturing block

The signal capturing module includes signal preprocessing and capturing brain signals through different invasive or noninvasive techniques such as electroencephalography (EEG): Sensors are affixed to the scalp to capture electrical impulses from the brain. Magnetoencephalography [11]: It measures the magnetic fields generated by brain activity using sensing devices. Functional magnetic resonance imaging (fMRI) [12]: It uses radio waves and magnetic fields to produce images of the brain's blood flow, making it possible to assess neural systems. Near-Infrared Spectroscopy (NIRS): This is a method that makes use of light to monitor fluctuations in neural activity by tracking variations in the amount of oxygenated and deoxygenated hemoglobin in the nervous system. In PET, radioactive tracers are injected into the body,

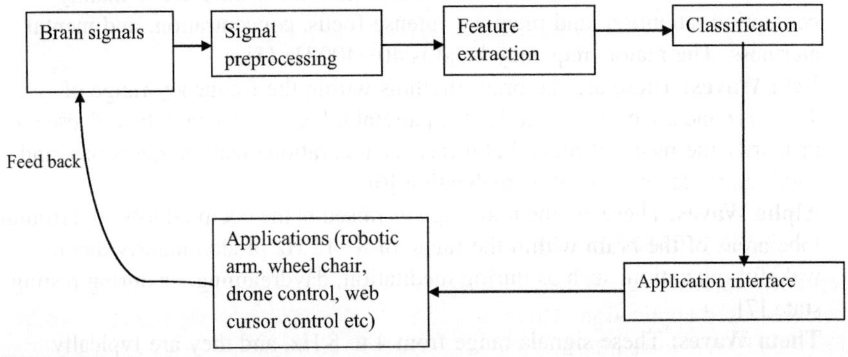

FIGURE 9.1

Block diagram of brain–computer interface system.

and the amount of tracer that reaches the brain is measured, allowing for the measurement of neural activity responsible for collecting the electrophysiological signals used by the BCI; these signals are obtained from the brain [13]. Electrocardiograms and single-neuron recordings are invasive methods that yield better results [13,14]. When compared with noninvasive electroencephalogram and magnetoencephalogram, positron emission tomography [15], and functional magnetic resonance imaging [16], near-infrared spectroscopy [17], and functional magnetic resonance imaging [14], signal strength is superior. The obtained signals are amplified to boost their strength before being transmitted. They must be encoded before any computer application.

1.2 Processing of signals block

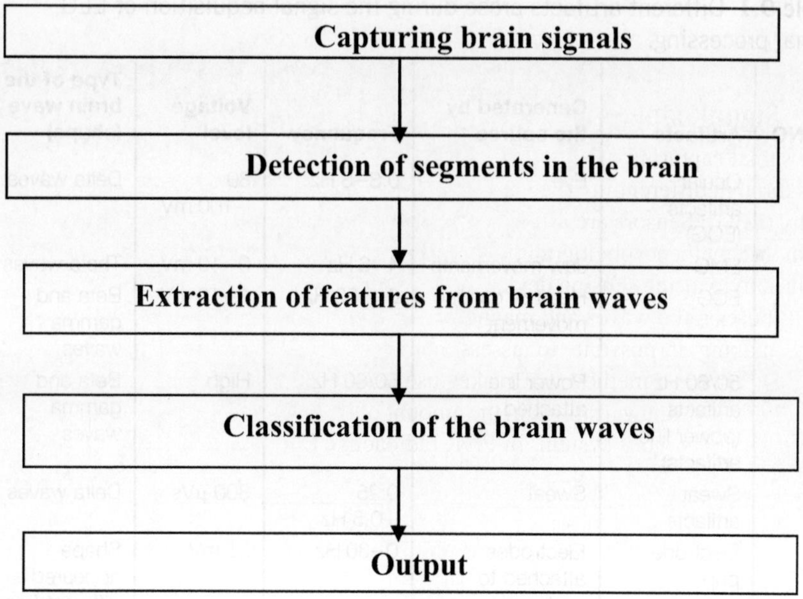

FIGURE 9.2

Processing of EEG signals.

1.3 Signal preprocessing

Preprocessing EEG signals shown in Fig. 9.2 is a critical step in any brain–computer interface-based application. This stage consists of removing artifacts such as ECG, EOG, and EMG, noise filtering, and signal resampling to meet sensor feedback. Preprocessing is the process of increasing the SNR of recorded data for preprocessing.

The element of the EEG signal induced by head muscle activity and eye gaze generates an electric action unconnected to brain function. This frequency component is thought to be an artifact and cannot be processed. To remove artifacts from EEG recordings, signal preprocessing is required. The proper preprocessing of the EEG signal is used by BCI systems. It helps to eliminate undesired artifacts from the EEG signal and prepares it for further processing. For example, Blind Source Separation (BSS) is the data preprocessing technique used [18].

The stutters in EEG signals are identified using various techniques such as eye blink detection, muscle movement detection, and power line interference detection. Following the detection of artifacts, various techniques such as independent component analysis, BSS, and principal component analysis are used to remove them [19]. Table 9.1 lists the various inaccuracies in EEG rhythms of various types. The existence of these artifacts can interact with EEG signal interpretation, resulting in

Table 9.1 Different artifacts arose during the signal acquisition of EEG signal processing.

S.NO	Artifacts	Generated by the source	Frequency	Voltage level	Type of the brain wave (shape)
1	Ocular artifacts (EOG)	Eye	0.3–3 Hz	80 −100 mV	Delta waves
2	EMG	Jaw movements	4–6 Hz	0–10 mV	Theta waves
3	ECG	Heart or cardiac movement	0–150 Hz	1–10 mV	Beta and gamma waves
4	50/60 Hz artifacts (power line artifacts)	Power line attached	50/60 Hz	High	Beta and gamma waves
5	Sweat artifacts	Sweat	0.25 −0.5 Hz	300 μVs	Delta waves
6	Electrode pop	Electrodes attached to scalp	0–30 Hz	20 mV	Shape appeared different from actual EEG signal
7	Physical movement artifacts (motion artifacts)	Body movements, head movement, jaw movement, etc.	Very low	High	Shape appeared different from actual EEG signal
8	Electronic gadgets artifacts	Mobile, laptop, personal computer, etc.	Very low	High	Shape appeared different from actual EEG signal

decreased performance and accuracy of BCI applications. As a result, it is critical to eliminate the artifacts before using the EEG signals for more analysis and processing. By removing the artifacts, the EEG signal quality can be improved, resulting in increased efficiency and accuracy in BCI systems.

1.4 Feature extraction

This is the method by which relevant attributes from EEG signals can be used to categorize the EEG signals. Power spectral density, time-frequency analysis, and spatial-temporal analysis are all methods for extracting features. These characteristics can be calculated using a variety of mathematical methods, including the Fourier transform and wavelet transform. It is critical to extract features that are relevant to the task at hand and are resistant to noise and artifacts [20].

1.5 Classification

This is the method for categorizing EEG signals into various mental states using feature extraction. Support vector machines, linear discriminate analysis, k-nearest neighbors, and deep learning algorithms [21] can all be used to accomplish this. The classifier chosen is determined by the task at hand as well as the computational resources available. It is critical to assess the classifier's efficiency using appropriate evaluation metrics such as accuracy, sensitivity, and specificity [22].

1.6 Application interface

This is the process of mapping the mental states to control commands. This can be done using a variety of techniques, such as the Kalman filter or other control algorithms. The choice of the application interface depends on the specific task and the available computational resources [23].

1.7 Applications

BCI systems have various uses, including controlling a cursor, a robotic arm, or a prosthetic limb in fields such as medicine, rehabilitation, gaming, and entertainment. Furthermore, in people with disabilities, BCI can be used to regulate and interact with outside devices such as a wheelchair or a smart home. BCI technique has also been used in neurofeedback, which is the process of providing the user with real-time information about brain activity to assist them in improving their mental state [24].

1.8 Feedback

The classification results are presented to the user in this step, providing feedback on the effectiveness of the BCI. The artifact's efficiency detection method can be validated using feedback. EEG signals can be compared with and without artifact removal to verify the signal improvement in quality. Feedback can be used to track the technique's efficiency for removing artifacts, for example, the degree to which the interference with the EEG data has been reduced. The technique's performance can be enhanced and optimized using this information. Finding the source of the artifacts: artifacts like eye movements, muscle contractions, or electrical noise can be found using feedback. By concentrating on the source of the artifacts, this information can be used to create more efficient artifact removal methods. Enhancing the technique's accuracy: The artifact removal technique can be made more accurate by using feedback. For instance, if the feedback shows that certain artifacts are not being used.

In general, it is possible to use the BCI overall system performance to gauge how well artifact removal techniques perform. The performance measures of BCI systems can be enhanced by reducing artifacts in EEG signals, which could result in more precise and dependable outcomes in application-specific tasks.

2. Artifacts removal techniques

Table 9.2 shows the comparison of the latest artifact removal techniques in various parameters such as the type of artifacts that can be eliminated in EEG signal processing which is mainly related to BCI applications, novelty in the algorithm or method chosen to mitigate artifacts, and the data that can be operated on which the proposed method can best suit (real and simulated) so that we can estimate practical implementation; it also discusses the challenges or limitations faced to practical viability and commented or given remarks about each and every system of implementation. Table 9.2 shows different artifact removal techniques such as EOG, ECG, physical movement artifacts (motion artifacts), etc., but it mainly focuses on ocular or eye blink (EB) artifacts because the EB artifacts are the main cause of error or distortion in EEG signal preprocessing.

Several strategies can be employed to eliminate artifacts from EEG signals in application fields such as BCI and EEG signal processing [43].

Independent component analysis: A process for disentangling reliable sources of EEG signals based on statistical traits. This technique can completely remove artifacts like eye blinks, muscle movements, and power line interference [44].
Principal component analysis (PCA): A strategy for removing noise and artifacts from EEG signals by projecting them onto a lower-dimensional space [45].
Empirical mode decomposition: A technique for decomposing EEG signals into a residue and a number of intrinsic modes (IMFs). This technique can completely remove nonstationary artifacts such as eye blinks, muscle movements, and power line interference [46].

Table 9.2 Comparison of the latest different artifact removal techniques.

Article	Type of artifact	Method	Algorithm	Novelty	Data	Challenges/ limitations	Ref.	Comments
Çınar, Salim (2021)	EOG	ICA, kurtosis, *k*-means,	LMS, NLMS, MZS, ANC	EOG signals must not necessitate an external electrode	Real and simulated	It is applicable only for ocular artifacts and is not applicable for other artifacts	[25]	This is applicable for all types of datasets and is a comfortable measurement for patients during more time EEG recordings
Trigui, Omar et al.	Eye blink	ADJAST, MARA, and SASICA	Genetic algorithm (GA)	Minimize the RMSE	Simulated	It is difficult to apply all types of datasets	[26]	The suggested method has high efficiency and can be employed with a tiny multitude of channels.
Cao, Jiuwen et al.	Only eye blink (EOG)	Gaussian mixture model (GMM)	GMM algorithm and cascaded hybrid threshold method	Correctly detected all EB artifacts without generating any false positive	Real and simulated	Many EB artifacts were missed because of using a high threshold.	[27]	The proposed algorithm is much more robust because it continually provides a higher precision and F1 score.

Continued

Table 9.2 Comparison of the latest different artifact removal techniques.—*cont'd*

Article	Type of artifact	Method	Algorithm	Novelty	Data	Challenges/ limitations	Ref.	Comments
Egambaram, Ashvaany et al.	Only eye blink (EOG)	Fast EMDCCA and Fast CCA	eADA with modified empirical decompossition and canonical correlation analysis	More than 97%removal accuracy and an average of 10–13 ms removal speed	Simulated	There is no significant distortion to be discovered	[28]	Feasible for removal of eye blink artifacts online with minimal neural distortion
Borowicz, Adam.	Only eye blink (EOG)	ICA and principle of regression analysis	Multichannel Wiener filter (MWF)	Modified ICA simpler than ICA	Real and simulated	Utilizing sophisticated multichannel linear filters,	[29]	The new technique is best suited to real-world systems than the current method.
Zhou, Weidong, and Jean Gotman	Only eye blink (EOG)	ICA method	The EEG dipole framework is combined with ICA.	Separating the EEG and noise-independent components without reliance on citation artifacts	Real and simulated	There is no difference between the frequency distributions of slow waves and eye artifacts.	[30]	This technique was found to be successful in removing the primary ocular artifacts from the EEG.
Sreeja, S. R. et al.	Mainly eye blink (EOG) and also used for other artifacts removal	Morphological component analysis (MCA) and K-SVD	Two sparsity-based techniques: MCA and K-SVD	There is no need for additional equipment for EB artifacts to be removed from the EEG signal using channels.	Real and simulated	Extraocular channels are required to capture eye artifacts.	[31]	Raw EEG signals can also be cleaned of other artifacts using this technique.

He, Ping, G. Wilson, and C. Russell	Eye blink artifacts	Adaptive filtering	Recursive least squares algorithm	This technique is used to track the nonstationary portions of the EOG signals.	Real	It is not a method that extended to cases with three or more reference inputs.	[32]	Adapt instantly to a new situation without losing efficiency
Chintala, Sridhar, and Jaisingh Thangaraj	Ocular artifacts	Robust variable forgetting factor (RVFF) and RLS	RVFF-RLS based algorithm	This algorithm acquires clean EEG signals by the subtraction	Real and simulated	Tracking performance deteriorates in nonstationary environments	[33]	In a challenging environment, the suggested methodology has the lowest mean square error.
Yadav, Anchal, and Mahipal Singh Choudhry.	Ocular artifacts	EEMD and SCICA Kurtosis and mMSE	EEMD and SCICA Kurtosis and mMSE	Additionally, EEMD is used to get around EMD's mode mixing and aliasing problems.	Real	EEMD amplitude reduced issue	[34]	The performance could be improved even further by using better ICA constraints and wavelet-enhanced independent component analysis.

Continued

Table 9.2 Comparison of the latest different artifact removal techniques.—*cont'd*

Article	Type of artifact	Method	Algorithm	Novelty	Data	Challenges/ limitations	Ref.	Comments
Gajbhiye, Pranjali, Rajesh Kumar Tripathy	Ocular artifacts	The rhythm separation technique on FBSE-EWT	The empirical wavelet transform based on the Fourier —Bessel series expansion	No reference signal is needed to filter out ocular artifacts from the EEG signal.	Real	Mode mixing occurs when EEG rhythm data is presented in multiple modes.	[35]	With fewer resources, the suggested method performs best. Compared to other methods currently in use,
Shi-Yun Shao et al.	Eye blink, ECG	ICA-weighted PWC-PSVM	PWC_PSVM with error correction	Modified probabilistic multiclass SVM	N/A	Not applicable to all datasets	[36]	ECG artifact, EOG artifact, and other artifacts can all be automatically removed.
Sourya Bhattacharyya et al.	All artifacts	Biclassification ICA	Burst pattern detection	Targets both transitory and monotonous artifact detection	N/A	Complexity in implementation	[37]	In newborns EEG biclassification is used to distinguish artifacts from bursts.
Xun Chen et al.	Muscle	EEMD-JBSS	EEMD-MCCA	Actually, realize single-channel EEG muscle artifact removal	Real and simulated	Limit standard deviation	[38]	The suggested approach is a good potential single-channel approach.

Islam, Md Kafiul, Parviz Ghorbanzadeh, and Amir Rastegarnia	All types of artifact removal (ECG, EOG, EMG, etc.)	Entropy, kurtosis, skewness, periodic waveform index	Stationary wavelet transform-based artifact removal	Results reveal that proposed artifact removal significantly improves BCI output	Real and simulated	The capability of removing artifacts is still limited	[39]	Proposed method maps artifacts probability based on four statistical measures which improve performance
Lee, Young-Eun, No-Sang Kwak, and Seong-Whan Lee	Movement artifacts	ICA with online learning	Online learning and constrained independent component analysis (cIOL)	Examine the impact of denoising on time and frequency domains	Real and simulated	Furthermore, artifacts are far too diverse to be represented by a single template.	[40]	EEG signals were used to approximate movement artifacts. The weights in each sample were then up to date using online learning, and artifact-free EEG signals were obtained.
Song, YoungJae, and Francisco Sepulveda	EMG artifacts	ICA, PCA, and BSS-CCA	EMG-CCh	Reduce uncertainty and enhance class separation	Simulated	Methodological restrictions in idle states, a channel with a lesser CRC may still have too much class-dependent EMG.	[41]	The approach can be utilized on its own or in conjunction with preexisting artifact handling methods.
AR Teixeira et al.	EOG, baseline drifts	Local SSA, ICA	Local singular spectrum analysis and ICA	Automatic removal of ocular, baseline artifacts	Real	Used for single-channel EEG measurement.	[42]	Both the cleaned-up EEG and the extracted artifacts will be displayed.

Wavelet-based methods: To eliminate noise and artifacts from EEG signals, wavelet-based methods like wavelet thresholding and demising can be used [47].

Adaptive filtering: To isolate artifacts from EEG signals, adaptive filtering methods like Kalman filters, Wiener filters, and RLS filters can be used [33]. Common spatial patterns (CSP) strategies can be employed in BCI to retrieve spatial patterns from EEG signals that are related to the underlying neural activity of interest while dismissing undesired artifacts [48].
To classify EEG signals into different states, machine learning techniques such as support vector machine and random forest [49] can be used to ignore artifacts, such as left or right arm movement [50].
Temporal filtering: In this method, high-frequency noise is subtracted from the EEG signal using a low-pass filter. By doing so, it may be possible to help eliminate artifacts caused by electrical noise, muscle activity, and eye movements [51].
Rejection-based methods: This method involves thresholding the data and removing data points that exceed the threshold. This is a simple method, but it may not be very effective in cases where the artifacts are nonstationary or occur at different times [52].
Common average reference: The EEG signals from all electrodes are averaged and then subtracted from the signals from each electrode in this spatial filtering method. This is simple and effective for lowering common artifacts [53].
Laplacian spatial filtering: This method is a spatial filtering method that calculates the Laplacian of the EEG signals, and it is obtained by subtracting the average of the neighboring electrodes' signals from each electrode's signal. This method is effective in reducing the artifacts caused by muscle movement [54].
Ensemble empirical mode decomposition: The input is broken down into a number of intrinsic modes (IMFs) using this technique, and the IMFs that correspond to the artifacts are then eliminated [46].
Optimal basis set: By minimizing the difference between the EEG signals and the linear combination, this technique is used to determine the EEG signals using a linear combination of basis functions [55].
Deep learning techniques: Recent research has suggested removing artifacts from EEG signals in BCI using deep learning techniques like convolutional neural networks or recurrent neural networks. Eye blinks, muscle movement, electrical noise, and other artifacts have all been found to be effectively eliminated by these techniques [56].
Generative adversarial networks: It has been suggested for use in BCI to remove artifacts from EEG signals by training a generator network [57] to produce artifact-free EEG signals and a discriminator network to distinguish between both the original signals and the artifact-free signals [58].

Hybrid methods: Various methods for trying to remove artifacts from EEG signals in BCI have been put forth in some research. For instance, it has been discovered that combining ICA with a deep learning-based approach can effectively get rid of a variety of artifacts [59].
Nonlinear noise cancellation: Nonlinear filters are used in this technique to clean up artifacts from EEG signals. The filters' purpose is to keep the preferred signals intact while suppressing artifacts [60].
Blind source separation (BSS): This technique separates the sources of the EEG signals. without revealing the mixing method or the number of alternatives. The artifacts are then eliminated using the separated sources [19].
Deep belief networks: In BCI, it has been suggested that artifacts from EEG signals be removed. using this kind of architecture for deep learning. DBNs have been demonstrated to be effective at eliminating a variety of artifacts, including electrical interference, muscle activity, and eye blink [61].
Temporal filtering, spatial filtering, independent component analysis, and rejection-based methods are some of the commonly used techniques for removing artifacts from EEG signals [62]. Although these techniques are straightforward to use, they may not be very effective in cases where the artifacts are nonstationary or occur at different times. Machine learning-based methods, such as SVM, Naive Bayes and other classifiers, have been EEG signals' artifacts have been successfully removed using this technique. These methods can be used to classify the data into artifact and nonartifact segments and then remove the artifact segments from the EEG signal.

Here are some ways to achieve this.

Artifact reduction techniques: Independent component analysis, wavelet transform, filtering, and regression analysis are a few methods that can be used to lessen artifacts in EEG signals.
Electrode placement: The placement of electrodes can have a significant impact on the quality of the use of CNN and RNN [63], two deep learning-based methods; It has also been suggested as a cutting-edge method to eliminate artifacts from EEG signals. Hybrid methods and generative adversarial networks have also been suggested as novel ways to clean up artifacts from EEG signals. To improve performance metrics in BCI applications, artifacts in EEG signals must be reduced, and the effectiveness of the EEG signals used for analysis must be improved [64]. Electrode placement should be carefully considered to make sure that they are placed in regions with strong EEG signals and that they are not impacted by muscle or eye movement artifacts.
Improved signal processing: The EEG signals used for the assessment can be made better by utilizing cutting-edge signal processing methods like adaptive filtering and machine learning techniques.
Multichannel recordings: EEG signals can be recorded with more electrodes using multichannel recordings, which can produce more precise and in-depth EEG data.

Depending on the scope and specifications of the BCI task, different performance measures may be selected in application-specific BCI. Typical performance metrics employed in a number of BCI applications include the following [65]: Event-related desynchronization [66], event-related synchronization [66], accuracy, and response time for motor imagery-based BCI. Recognition rate, accuracy, For P300-based BCI, the false positive rate and reaction time. For the Steady State, recognition rate, precision, rate of false positives and response time for Visually Evoked Potential BCI. Negativity resulting from errors (ERN) BCI: accuracy, reaction time, and ERN amplitude and latency. Hybrid BCI: Combining a variety of metrics from various BCI applications based on the job at hand.

Depending on the particular requirements and constraints of the task, artifact reduction techniques' effects can change in application-specific tasks. Effective artifact removal, for instance, can lead to increased accuracy in identifying the intended motor imagery movements in a BCI for motor imagery. Efficient artifact removal can improve the target stimulus detection accuracy and response time in a P300-based BCI. Therefore choosing an artifact removal technique that is appropriate for the specific requirements of the BCI application is crucial. This could aid in enhancing the BCI system's final outcomes and performance metrics.

Methods for removing artifacts are essential in BCI systems because they help to increase the consistency and precision of the recorded EEG signals. The type of artifact, the application, and the desired result all influence the choice of artifact removal techniques. Signal-to-noise ratio (SNR), correlation, and classification accuracy are a few examples of performance indicators used in the artifact removal method.

From Table 9.2, the main methods to remove artifacts in EEG signal processing are blind source separation, discrete wavelet transform, stationary wavelet transform, empirical mode decomposition, ensemble empirical mode decomposition, wavelet transform, and adaptive filtering that worked on different methods to remove the online artifacts for real-time applications, which are mainly used for EEG signal processing; further, it is applied mainly to advance technologies like brain—computer interface applications. When preprocessing stage improved, then the features and classification of signals automatically increase the performance parameters such as correlation coefficient, mean square error, power spectral density, signal-to-noise ratio, accuracy, execution speed, and complexity [67].

Summary of Table 9.2 discusses advanced artifact removal techniques for the case studies illustrated, such as nar, Çınar [68], who dissected and implemented the new algorithms using the methods of kurtosis, *k*-means, modified Z-score, and adaptive noise canceler. Independent component analysis is a technique [69] that can be used with both real and simulated data and is relevant only for eye blink artifacts., the proposed system does not necessitate the use of an external electrode for measurement EOG signals. The drawback of the system is it is only applicable to ocular artifacts and other artifacts present it is not efficient method and the drawback of this method is that the subtraction step may remove the proper EEG signals, and the other paper Borowicz, Adam discussed about the independent component analysis (ICA) and principles of regression analysis and implemented. The proposed algorithm is easier in PCA than the ICA approach, as it employs a multichannel Wiener filter and a subset of the frontal electrodes. It is also well-suited to practical cases, which appears to be crucial in the advancement of BCI. Zhou, Weidong, and Jean Gotman also implemented a new proposal blending the EEG dipole with the independent component analysis model; this method also mainly concentrated on the removal of ocular artifacts only. This method was discovered to be successful in automatically removing ocular artifacts embedded in the EEG. Lee, Young-Eun, No-Sang Kwak, and Seong-Whan Lee discussed about the ICA with online learning and artifact disposal, with a focus on motion artifacts. This is somewhat different from what was mentioned above. Song, YoungJae, and Francisco Sepulveda also implemented the system using ICA, along with PCA, and BSS-CCA to remove EMG artifacts by a novel technique called EMG-cch and is best suited for use along with other techniques the data only implemented on simulation results.

Trigui, Omar, and colleagues proposed a genetic algorithm (GA) method that decreases the root mean square error between raw and cleaned EEG. The proposed approach works well and can be used with a small number of channels and only simulated data.

Cao, Jiuwen et al. proposed a method of Gaussian mixture model (GMM) [27]; all eye blink artifacts were successfully recognized by the suggested technique, with no false positives.

Egambaram, Ashvaany, and colleagues used a method; Fast EMD-CCA and Fast CCA present a modified empirical mode decomposition and canonical correlation analysis-based unsupervised eye blink artifact detection method [70] with over 97% removal accuracy and an average removal speed of 10−13 ms. Online eye blink artifact removal using low neural deviation.

The methodologies proposed by Sreeja, S. R. et al. include morphological component analysis, the K-SVD-based artifact reducing method, and sparsity-based methods, which work on both real and simulated data. No additional equipment or EOG channels, parameter tinkering like thresholding, or channel information is needed to remove EB artifacts from the EEG signal. This technology's main flaw is that it needs extraocular channels to record ocular artifacts. Additionally, it can be used to remove extraneous artifacts from raw EEG data.

He, Ping, G. Wilson, and C. Russell presented adaptive filtering, utilizing the recursive least squares technique for ocular artifacts. On real-world data, this approach tracks the nonstationary part of the EOG signals. It is not a method that can be expanded to cases with three or more reference inputs, but it can be automatically modified to a new circumstance without losing its effectiveness.

Chintala, Sridhar, and Jaisingh Thangaraj focused solely on ocular artifacts and developed a solution based on the robust variable forgetting factor (RVFF) and recursive least square (RLS) [71]. This technique tracks and estimates nonstationary EOG signals to extract clean EEG signals from real and simulated data. Tracking performance suffers in nonstationary environments. In a dynamic setting, the proposed approach has the lowest mean square error.

Yadav, Anchal, and Mahipal Singh Choudhry remove ocular artifacts using EEMD and SCICA computation of Kurtosis and mMSE, and the algorithm used is ensemble empirical mode decomposition and spatial constraint independent component analysis in which EEMD [39] is also used to overcome the mode mixing and aliasing problem of EMD primarily performed on real data; the disadvantage of this technique is EEMD amplifier. The performance can be increased further by implementing better constraints on ICA and wavelet-enhanced independent component analysis. Gajbhiye, Pranjali, and Rajesh Kumar Tripathy proposed an ocular artifact can be eliminated using the FBSE-EWT-based separation technique [40]. The Fourier–Bessel series expansion-based empirical wavelet transform approach [35] was tested primarily on real-valued data and does not require a reference voltage for the filtering of ocular artifacts from an EEG signal. When the EEG rhythm information manifests in a variety of modes, mode mixing occurs. The method outperforms other current methodologies, with an average minimum MAE in PSD value for rhythm of 0.029.

Islam, Md Kafiul, Parviz Ghorbanzadeh, and Amir Rastegarnia [41] suggested a method for removing all types of artifacts (ECG, EOG, EMG, etc.) based on the concepts of entropy, kurtosis, skewness, and stationary wavelet transform. The results show that the suggested artifact removal greatly enhances BCI output when tested on real and simulated data. The proposed method's discrimination capacity still has to be enhanced, and its ability to remove genuine artifacts is poor. Based on four statistical parameters, the proposed method maps artifactual probability.

3. Conclusion

The work is mostly considered in the preprocessing step of the overall BCI systems. The goal of the preprocessing stage in BCI applications is to decrease artifacts in the EEG signal generated by the numerous sources. Based on the findings in the available literature, this report summarized the key techniques. The techniques uses exclusively used for removing artifacts which are related to eye blink (EOG) artifacts, ECG, EMG and all other movement-related artifacts hereby go through the different research articles basically uses different algorithms separately or combine that reveals the output without artifacts in EEG signal processing which combined with BCI related applications either it may be cursor movement, wheelchair movement, video gaming, biomedical, etc. Some methods, such as adaptive filtering, morphological component analysis (MCA), and K-SVD and entropy, kurtosis, skewness, and periodic waveform index, remove artifacts with high precision, which works on both real and simulated data or either of the one method with high computational costs, however, might not be appropriate for online applications. There is therefore no ideal method for getting rid of all kinds of artifacts. Therefore, the long-term objective of effective artifact attenuation is to offer a framework methodology with enhanced precision, time, and accuracy.

References

[1] Kübler A. The history of BCI: from a vision for the future to real support for personhood in people with locked-in syndrome. Neuroethics 2020;13(2):163–80.

[2] Kawala-Janik A. Efficiency evaluation of external environments control using bio-signals. Ph.D. Thesis. London, UK: University of Greenwich; 2013.

[3] Ebersole JS, Pedley TA. Current practice of clinical electroencephalography. Philadelphia, PA, USA: Lippincott Williams & Wilkins; 2003.

[4] Millett DHB. From psychic energy to the EEG. Perspectives in Biology and Medicine 2001;44:522–42.

[5] Vialatte FB, et al. EEG paroxysmal gamma waves during Bhramari Pranayama: a yoga breathing technique. Consciousness and Cognition 2009;18(4):977–88.

[6] Inuso G, et al. Wavelet-ICA methodology for efficient artifact removal from Electroencephalographic recordings. In: 2007 international joint conference on neural networks. IEEE; 2007.

[7] Fouad MM, et al. Brain computer interface: a review. Brain-Computer Interfaces 2015: 3–30.

[8] Kim KH, Yoon HW, Park HW. Improved ballistocardiac artifact removal from the electroencephalogram recorded in fMRI. Journal of Neuroscience Methods 2004;135(1–2): 193–203.

[9] Tan D, Nijholt A. Brain-computer interfaces and human-computer interaction. Springer London; 2010.

[10] Alimadadi A, et al. Artificial intelligence and machine learning to fight COVID-19. Physiological Genomics 2020;52(4):200–2.

[11] Stam CJ. Use of magnetoencephalography (MEG) to study functional brain networks in neurodegenerative disorders. Journal of the Neurological Sciences 2010;289(1–2): 128–34.

[12] DeYoe EA, et al. Functional magnetic resonance imaging (FMRI) of the human brain. Journal of Neuroscience Methods 1994;54(2):171–87.

[13] Donoghue JP. Connecting cortex to machines: recent advances in brain interfaces. Nature Neuroscience 2002;5:1085.

[14] Serruya Mijail D, et al. Brain-machine interface: instant neural control of a movement signal. Nature 2002;416:141.

[15] Muehllehner G, Karp JS. Positron emission tomography. Physics in Medicine and Biology 2006;51(13):R117.

[16] Matthews PM, Peter J. Functional magnetic resonance imaging. Journal of Neurology, Neurosurgery & Psychiatry 2004;75(1):6–12.

[17] Siesler HW, et al., editors. Near-infrared spectroscopy: principles, instruments, applications. John Wiley & Sons; 2008.

[18] Cichocki A, et al. EEG filtering based on blind source separation (BSS) for early detection of Alzheimer's disease. Clinical Neurophysiology 2005;116(3):729–37.

[19] Jung T-P, et al. Removing electroencephalographic artifacts by blind source separation. Psychophysiology 2000;37(2):163–78.

[20] Vega-Escobar L, Castro-Ospina AE, Duque-Muñoz L. Feature extraction schemes for BCI systems. In: 2015 20th symposium on signal processing, images and computer vision (STSIVA). IEEE; 2015.

[21] Al-Dabagh MZN, Alhabib MHM, AL-Mukhtar FH. Face recognition system based on kernel discriminant analysis, k-nearest neighbor and support vector machine. International Journal of Religious Education 2018;5(3):335–8.
[22] Blumberg J, et al. Adaptive classification for brain computer interfaces. In: 2007 29th annual international conference of the IEEE engineering in medicine and biology society. IEEE; 2007.
[23] Zander TO, Kothe C. Towards passive brain–computer interfaces: applying brain–computer interface technology to human–machine systems in general. Journal of Neural Engineering 2011;8(2):025005.
[24] Moore MM. Real-world applications for brain-computer interface technology. IEEE Transactions on Neural Systems and Rehabilitation Engineering 2003;11(2):162–5.
[25] Çınar S. Design of an automatic hybrid system for removal of eye-blink artifacts from EEG recordings. Biomedical Signal Processing and Control 2021;67:102543.
[26] Trigui O, et al. Removal of eye blink artifacts from EEG signal using morphological modeling and orthogonal projection. Signal, Image and Video Processing 2022;16(1):19–27.
[27] Cao J, et al. Unsupervised eye blink artifact detection from EEG with Gaussian mixture model. IEEE Journal of Biomedical and Health Informatics 2021;25(8):2895–905.
[28] Wang J, et al. Eye blink artifact detection with novel optimized multi-dimensional electroencephalogram features. IEEE Transactions on Neural Systems and Rehabilitation Engineering 2021;29:1494–503.
[29] Egambaram A, et al. Online detection and removal of eye blink artifacts from electroencephalogram. Biomedical Signal Processing and Control 2021;69:102887.
[30] Borowicz A. Using a multichannel Wiener filter to remove eye-blink artifacts from EEG data. Biomedical Signal Processing and Control 2018;45:246–55.
[31] Zhou W, Gotman J. Automatic removal of eye movement artifacts from the EEG using ICA and the dipole model. Progress in Natural Science 2009;19(9):1165–70.
[32] Sreeja SR, et al. Removal of eye blink artifacts from EEG signals using sparsity. IEEE Journal of Biomedical and Health Informatics 2017;22(5):1362–72.
[33] He P, Wilson G, Russell C. Removal of ocular artifacts from electro-encephalogram by adaptive filtering. Medical, & Biological Engineering & Computing 2004;42(3):407–12.
[34] Joyce CA, Gorodnitsky IF, Kutas M. Automatic removal of eye movement and blink artifacts from EEG data using blind component separation. Psychophysiology 2004;41(2):313–25.
[35] Chintala S, Thangaraj J. Ocular artifact elimination from EEG signals using RVFF-RLS adaptive algorithm. In: 2020 national conference on communications (NCC). IEEE; 2020.
[36] Shao S-Y, et al. Automatic EEG artifact removal: a weighted support vector machine approach with error correction. IEEE Transactions on Biomedical Engineering 2008;56(2):336–44.
[37] Bhattacharyya S, et al. Detection of artifacts from high energy bursts in neonatal EEG. Computers in Biology and Medicine 2013;43(11):1804–14.
[38] Chen X, Chen H, Hu P. Removal of muscle artifacts from single-channel EEG based on ensemble empirical mode decomposition and multiset canonical correlation analysis. Journal of Applied Mathematics 2014;2014.
[39] Yadav A, Singh Choudhry M. A new approach for ocular artifact removal from EEG signal using EEMD and SCICA. Cogent Engineering 2020;7(1):1835146.

[40] Gajbhiye P, Kumar Tripathy R, Pachori B. Elimination of ocular artifacts from single channel EEG signals using FBSE-EWT based rhythms. IEEE Sensors Journal 2019; 20(7):3687–96.

[41] Islam MK, Ghorbanzadeh P, Rastegarnia A. Probability mapping based artifact detection and removal from single-channel EEG signals for brain–computer interface applications. Journal of Neuroscience Methods 2021;360:109249.

[42] Teixeira AR, et al. Automatic removal of high-amplitude artefacts from single-channel electroencephalograms. Computer Methods and Programs in Biomedicine 2006;83(2): 125–38.

[43] Vaid S, Singh P, Kaur C. EEG signal analysis for BCI interface: a review. In: 2015 fifth international conference on advanced computing & communication technologies. IEEE; 2015.

[44] Barbati G, et al. Optimization of an independent component analysis approach for artifact identification and removal in magnetoencephalographic signals. Clinical Neurophysiology 2004;115(5):1220–32.

[45] Jung T-P, et al. Removing electroencephalographic artifacts: comparison between ICA and PCA. In: Neural networks for signal processing VIII. Proceedings of the 1998 IEEE signal processing society workshop (Cat. No. 98TH8378). IEEE; 1998.

[46] Sweeney KT, McLoone SF, Ward TE. The use of ensemble empirical mode decomposition with canonical correlation analysis as a novel artifact removal technique. IEEE Transactions on Biomedical Engineering 2012;60(1):97–105.

[47] Molavi B, Dumont GA. Wavelet-based motion artifact removal for functional near-infrared spectroscopy. Physiological Measurement 2012;33(2):259.

[48] Li D, et al. A self-adaptive frequency selection common spatial pattern and least squares twin support vector machine for motor imagery electroencephalography recognition. Biomedical Signal Processing and Control 2018;41:222–32.

[49] Ahmad I, et al. Performance comparison of support vector machine, random forest, and extreme learning machine for intrusion detection. IEEE Access 2018;6:33789–95.

[50] Radüntz T, et al. Automated EEG artifact elimination by applying machine learning algorithms to ICA-based features. Journal of Neural Engineering 2017;14(4):046004.

[51] Kellman P, Epstein FH, McVeigh ER. Adaptive sensitivity encoding incorporating temporal filtering (TSENSE). Magnetic Resonance in Medicine 2001;45(5):846–52.

[52] Jung T-P, et al. Removal of eye activity artifacts from visual event-related potentials in normal and clinical subjects. Clinical Neurophysiology 2000;111(10):1745–58.

[53] Foodeh R, et al. Minimum noise estimate filter: a novel automated artifacts removal method for field potentials. IEEE Transactions on Neural Systems and Rehabilitation Engineering 2016;25(8):1143–52.

[54] Li W, et al. Integrated Laplacian-based phase unwrapping and background phase removal for quantitative susceptibility mapping. NMR in Biomedicine 2014;27(2): 219–27.

[55] Niazy RK, et al. Removal of FMRI environment artifacts from EEG data using optimal basis sets. NeuroImage 2005;28(3):720–37.

[56] Ghani MU, Clem Karl W. Deep learning based sinogram correction for metal artifact reduction. Electronic Imaging 2018;2018(15):472–81.

[57] Sawangjai P, et al. EEGANet: removal of ocular artifacts from the EEG signal using generative adversarial networks. IEEE Journal of Biomedical and Health Informatics 2021;26(10):4913–24.

[58] Wang J, et al. Conditional generative adversarial networks for metal artifact reduction in CT images of the ear. Proceedings, Part I. In: Medical image computing and computer assisted intervention—MICCAI 2018: 21st international conference. Granada, Spain: Springer International Publishing; September 16—20, 2018. 2018.
[59] Sheela P, Puthankattil SD. A hybrid method for artifact removal of visual evoked EEG. Journal of Neuroscience Methods 2020;336:108638.
[60] Renevey P, et al. Wrist-located pulse detection using IR signals, activity and nonlinear artifact cancellation. In: 2001 conference proceedings of the 23rd annual international conference of the IEEE engineering in medicine and biology society. Vol 3. IEEE; 2001.
[61] Son Y, et al. Automated artifact elimination of physiological signals using a deep belief network: an application for continuously measured arterial blood pressure waveforms. Information Sciences 2018;456:145—58.
[62] Delorme A, Scott M, Sejnowski T. Automatic artifact rejection for EEG data using high-order statistics and independent component analysis. In: Proceedings of the third international ICA conference; 2001. p. 9—12.
[63] Baek S-S, Pyo J, Chun JA. Prediction of water level and water quality using a CNN-LSTM combined deep learning approach. Water 2020;12(12):3399.
[64] Usakli AB. Improvement of EEG signal acquisition: an electrical aspect for state of the art of front end. Computational Intelligence and Neuroscience 2010;2010.
[65] Kubler A, et al. BCI meeting 2005-workshop on clinical issues and applications. IEEE Transactions on Neural Systems and Rehabilitation Engineering 2006;14(2):131—4.
[66] Pfurtscheller G. EEG event-related desynchronization (ERD) and synchronization (ERS). Electroencephalography and Clinical Neurophysiology 1997;1(103):26.
[67] Sweeney KT, et al. A methodology for validating artifact removal techniques for physiological signals. IEEE Transactions on Information Technology in Biomedicine 2012; 16(5):918—26.
[68] Çınar S. Design of an automatic hybrid system for removal of eye-blink artifacts from EEG recordings. Biomedical Signal Processing and Control 2021;67:102543.
[69] Pandey KK, Shukla D. Min—max kurtosis mean distance based k-means initial centroid initialization method for big genomic data clustering. Evolutionary Intelligence 2022: 1—22.
[70] Egambaram A. Détection et suppression en temps réel des artefacts de clignement de l'œil dans les signaux issus d'un électroencéphalogramme. Diss. Bourgogne Franche-Comté 2020.
[71] Drašković S, Đurović Ž, Vera P. Absolute finite differences based variable forgetting factor RLS algorithm. IET Signal Processing 2022;16(1):80—91.

Further reading

[1] Chapter 2 - technological basics of EEG recording and operation of apparatus. In: Abhang PA, Gawali BW, Mehrotra SC, editors. Introduction to EEG- and speech-based emotion recognition. Academic Press; 2016.
[2] Aggarwal S, Chugh N. Signal processing techniques for motor imagery brain computer interface: a review. Array 2019;1:100003.

CHAPTER

Deep learning system of naturalistic communication in brain—computer interface for quadriplegic patient

10

N. Yuvaraj, K. Mouthami, Gitanjali Wadhwa, Srivatsa Sundarraj and S.A. Srinivasan

Department of Computer Science and Engineering, KPR Institute of Engineering and Technology, Coimbatore, Tamil Nadu, India

1. Introduction

The ability to identify whether an eye is open or closed is crucial in a variety of contexts, including the human—device interface and driver drowsiness monitoring systems. In medicine, keeping track of the patient's eye movement is vital. On an electroencephalogram (EEG) or ocular magnetic resonance imaging (MRI) [1,2], eye blinking can produce an artifact. Additionally, people who are severely incapacitated can communicate by blinking [3]. Quadriplegic individuals typically utilize their voices to locate the caretaker. However, due to lengthy ventilator care or challenging sputum excretion, aggressive quadriplegic patients are typically tracheostomized. In a tracheostomy condition, the patient is unable to speak. For these patients, calling for help requires alternative channels of communication.

A deep neural network (DNN) with three or multiple layers [4]. These neural networks try to mimic function of the human brain and allow it to "knowledge" from enormous amounts of data, even though they cannot match it in any way. More hidden layers can help improve and optimize accuracy, even if a neural network with only one layer may still provide approximations.

Recently, artificial intelligence (AI) has achieved some very impressive strides [5]. The use of AI and deep learning algorithms (DLAs) for image recognition in numerous medical domains has grown [6]. These domains comprise histology classifications, endoscopic discoveries, and radiographic imaging like X-ray or MRI scans [7,8]. Some studies have reportedly used these DLA to track the patient's eye movement. Using a DLA, we developed a novel approach for monitoring eye movement [9]. Additionally, we used our algorithm to create a method allowing quadriplegic patients to contact their careers.

There would be numerous signals close to the head, including the teeth, heart, eyes, and facial muscles. Hence, other eye movement sources frequently affect scalp

Computational Intelligence and Deep Learning Methods for Neuro-rehabilitation Applications
https://doi.org/10.1016/B978-0-443-13772-3.00009-1

EEG. Artifacts, or signals from sources other than the brain, that are diverse in the signal produced in the brain and obstruct a precise EEG inspection. Due to this, most EEG investigations use a preprocessing method to get artifact-free EEG signals.

Eye blinking is one of the most significant sources of artifacts tainting EEG recordings. Eye blink artifacts invariably contaminate EEGs in most EEG paradigms involving visual stimuli or visual feedback because natural blinking is an autonomic process of the human body [10]. Eyeblink has shown that the altered alpha and beta band EEG readings are considerable; however, theta band is most severely damaged when such artifacts are present. Cleaning EEG signals to get correct EEG analysis results, it is crucial to preprocess data affected by eye blink artifacts. The primary method for cleaning EEG data involves rejecting the artifact-contaminated periods or recovering the artifact-free data, and both strategies frequently call for an artifact identification procedure. However, identifying contaminated epochs is essential for the rejection approach [11]. By applying the recovery technique just to the detected areas or employing various correction factors, eye blink artifacts, and other muscle movement artifacts for the detected regions, the recovery strategy additionally makes use of the artifact identification procedure to reduce source data distortion.

The ability to identify artifacts is also useful for independent component analysis (ICA), which analyses EEG data using separate components that artifacts have tainted. Removal of Numerous earlier investigations on the rapid recognition of eye blink artifacts would be conducted; the works could be loosely differentiated into two groups: feature-based techniques and distance-based techniques. Feature-based techniques use features that have been retrieved following preprocessing steps like band-pass filtering [12,13] and component separation to detect the presence of artifacts in an epoch. Many new features have been included, including the Teager—Kaiser energy operator, kurtosis, entropy, and maximal absolute value. A template in distance-based methods represents the target artifact's general shape. The length between the design and a specific portion of the signal is computed after the template has been constructed by an expert or formed by averaging sample artifact signals. The distance has been calculated using several techniques, including cross-correlation, dynamic temporal warping (DTW), support vector machines (SVM), and correlation between EEGs in various channels. To determine whether or not artifacts taint a specific epoch, both techniques for detecting artifacts generally need the setting of a threshold value. The simplest method is to utilize a fixed threshold value derived experimentally from earlier investigations.

However, because the patterns and signals of the eye blink distortions vary widely and independently, using the same threshold value might cause severe detection mistakes in some people's EEGs [14]. As a result, the threshold values must be adjusted for each person's EEG. The authors developed a technique that makes use of the variation in the distribution of artifacts; however, to calculate the variance in the distribution of artifacts, they used expert intervention. According to their proposal, a training strategy necessitates using a secondary dataset with all artifacts that are manually classified [15]. The proportion of artifact-free data should be used to determine a threshold using a semisupervised approach, which was also advised. This technique falls under the category of a semisupervised method because this technique uses labeled, artifact-free EEG data instead of blink-labeled EEG data to estimate the distribution of standard data. In many end-user applications, real-time EEG processing is frequently required, making the deployment of these supervised algorithms difficult.

2. Related works

Some techniques help patients to communicate with the people around them. The deep learning techniques help patients who can use only their eye movements as a form of contact. Paralyzed patients are becoming more prevalent every day, and adequately supporting them requires 24 × 7 monitoring. Amyotrophic lateral sclerosis, brainstem lesions, stupor, and muscular dystrophy are a few of the neurodegenerative conditions that have an impact on how the body's muscles work. The affected people could not relocate and do their regular work. They require some assistive technology that uses biosignals to help them solve their difficulty. The famous scientist and theoretical physicist Stephen Hawking had amyotrophic lateral sclerosis and became paralyzed. He used his cheek muscles and blinking as an infrared switch by which he could scan and control the characters on the screen. Through this, he could browse the Internet, send emails, and speak through a voice synthesizer. Such Human-to-computer interaction (HCI) based on electrooculography (EOG) is one method utilized recently to solve this issue. In this study, we thoroughly examine the viability of realizing nine states of HCI using the way we provide. Five electrodes were implanted on the subject's face, close to their eyes, to collect the signals. These signals were processed using source power and band power approach after being amplified by the ADT26 bioamplifier, filtered, and processed.

The system can differentiate between opened and closed eyes because it was specified well enough to identify the eye characteristics. The suggested system can properly detect the eye condition at a rate of high, which is better than previous algorithms, which are similar, according to experimental data from test pictures [16]. Eye-tracking computer systems (ETCSs) operate by illuminating the surface of the eye with infrared light that is undetectable to the human eye. The gadget can calibrate and locate whether the user is looking at the camera that captures the pupil center's location and the cornea's reflection pattern. As a result, they may utilize a computer system that serves several purposes, "type" text, browse the Internet, and use communication software. ETCSs are considered "electronic aids to daily life" as they allow users to communicate and manage their surroundings [17,18].

According to certain research, TensorFlow, a high-level machine learning framework, was employed to support the infrastructure requirements. A publicly accessible neural network (inception), initially created for computer vision applications, is retrained using straightforward time series plots produced directly from accelerometer and gyroscope information [19]. However, their method could have been more suitable for real-time processing due to the lengthy maximization process. They used the threshold value of the 80th percentile of the specific data distribution and the thresholding approach initially described for picture binarization. However, both techniques were still reliant on empirical parameterization [20]. The process of the model suggests a novel method for the unsupervised real-time identification of eye blink artifacts, a novel technique for measuring fear based on nonintrusive data collected from several sensors. According to experimental findings based on the result size, *t*-test, and the sum of the connections between worths with other modal quality, the evaluation of fear using facial temperature and subjective judgment is more accurate than using an electroencephalogram (EEG) and the rate at which one blinks their eyes [21]. The suggested strategy's effectiveness is assessed in real-time and simulated settings. We attempted to show the following throughout the validation process.

(1) If the usage of an independent threshold improves the exactness of identifying eye blink artifacts in contrast to the frequent threshold;
(2) The efficiency of the real-time adaption algorithm; and
(3) How we may use the information to increase the recognition of eye blink artifacts' accuracy (Table 10.1).

3. Dataset

Image data collection was intended for communication with caregivers and can be manipulated based on eye movements. It can be controlled by the direction given by the eye. The image dataset consists of four folders and comprises a total of 14,500 images. The images in the dataset are regularized and augmented to achieve high accuracy in the first training session. The image dataset is named "Eye-dataset," which contains four directories. These four directories contain different eye images: the closed look of the eye, the forward look of vision, the right look of an eye, and the left look of an eye [22,23]. Here, the collection methodology of the dataset is different: at first, collected a few images; then, edited those images to achieve the required position and angle.

The image data mentioned explain the two kinds of data: training data and testing data. These data are separated into a ratio of 4:1 and is shown in Table 10.2. The training data explain the data which help train the data, and the testing data help test the data, which help build the complete model.

The different types of eye images are used in the dataset to implement the model, which helps the caretakers know what the patients were trying to communicate with them. Four types of images help in processing the data, and they are eyes with a closed look, eyes with an open look, eyes with a left look, and eyes with a right look (Fig. 10.1).

Table 10.1 Literature survey.

Authors	Technique	Performance	Advantage	Disadvantage
Wezel, Joep, Anders Garpebring	MR scanning machine is used	High-resolution images of eyes are collected	Data can be generated	Duplicate eye images can be created
Choi, Jong-Suk, Jae Won Bang, Hwan Heo	Image recognition	Eyeblink monitorization	Data can be created for implementation	Data quality is reduced at night time
Veiga, Jose Juan Dominguez, Martin O'Reilly	Convolutional neural network is used	Capture images of eyes from the patients to generate data	High accuracy in the results	Time consumption
Esteva, Andre, Brett Kuprel	Deep learning algorithms are used	Image recognition By labeled data	It can be used on smartphones	Accuracy is low when multiple classes are used
Feng, Rui, Marcus Badgeley, J. Mocco	Probability-based distribution technique	The ability of the body has resulted	Rapid results are generated	High cost for implement
Hirasawa, Toshiaki, Kazuharu Aoyama, Tetsuya Tanimoto	Single-shot multibox detection techniques are used	It helps to detect cancer cells in the stomach	Processed using a large dataset	False positive is high. Hence, accuracy is low
Kim KW, Hong HG, Nam GP, Park KR	In deep learning algorithms, the SVM model is used	Effects of the eye are detected using the image recognition	Eye issues can be detected	Cannot detect the eye image if it is half closed
Krizhevsky, Alex, Ilya Sutskever	Deep CNN techniques are used	1.2 million high-resolution images are classified using CNN	High accuracy only with supervised data	Usual algorithms are not helpful for large datasets.
Hoffmann, Sven, and Michael Falkenstein	Independent component analysis	Prediction of eye blink an eye movement of a person	The component-based method is more useful	Complexity and data requirement is high

Continued

Table 10.1 Literature survey.—*cont'd*

Authors	Technique	Performance	Advantage	Disadvantage
Lugo, Zulay, Fréderic Pellas, Veronique Blandin	Psychological impact and quality of life	30relationsof patients with LIS completed the FNQ and analyzed it	Percentage of satisfaction is acquired	Anxiety feelings and depressive thoughts
Nakayama, Yuki, Toshio Shimizu, Yoko Mochizuki	Tracheostomy-invasive ventilation (TIV)	Total quadriplegia is a predictor of severe communication impairment	Best methods of communication	Highly recommend early detection of impaired communication.
Olczak, Jakub, Niklas Fahlberg, Atsuto Maki	Image classification using deep learning algorithms	Determine the feasibility of skeletal radiographs using deep learning	VGG16 had the best results	The process of creating raw text from which labels must be removed
Tafreshi, Marzieh, and Ali Mohammad Fotouhi.	Support vector classifier	Local maximum vertical derivative patterns in eye areas are extracted	Eye-opening or closing detection	High computational complexity
Van Middendorp, J. J., F. Watkins, C. Park	Eye-tracking computer system	A camera captures the cornea's reflection pattern and the pupil's center's location	High accuracy obtained	Duplicate images will be barely recognized
Verleger, Rolf.	High-pass filtering	Most blinks were detected	This effect occurred not only in model	Bias results when records are made only using a short time constant.
Chang, Won-Du, Ho-Seung Cha, Kiwoong Kim	Eyeblink artifact detection using SDW	An essential step toward raising the standard of EEG data	Useful for headband-type wearable EEG applications	The high complexity of EEG data
Pult, Heiko, Britta H. Riede-Pult, and Paul J. Murphy	Digital motion tracking	Counted the number of blinks required to remove the LG	The blink frequency was higher in female than in male	Incomplete blinking occurred significantly less often in women than in men but was not related to age

Table 10.2 Data processing ratio.

Training data	80%
Testing data	20%

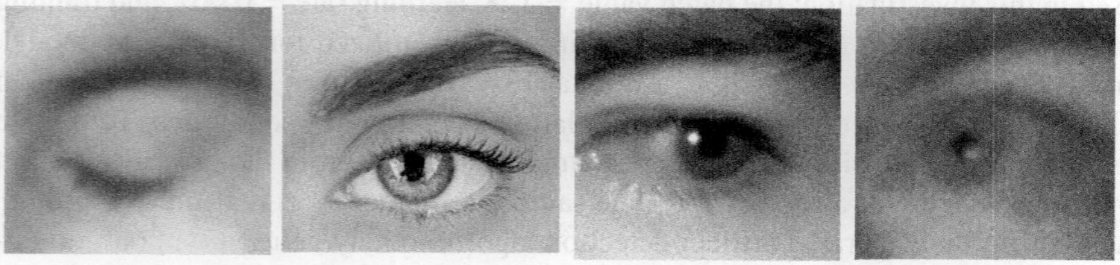

FIGURE 10.1

Dataset images. (A) Closed look of eye, (B) forward look of eye, (C) left look of eye, and (D) right look of eye. These images are collected and categorized into subfiles of the dataset. These images are used for processing the model.

4. Methodology

The images of eyes were recognized using computer vision (CV). CV is an open-source platform to work with images and videos. The program runs, after which it extracts the pictures of both eyes and identifies the face. The camera continually records and retains 15 eye pictures every second. The photos we gathered were 28×28 pixels, for a total of 784 pixels. In this investigation, eye scans from 30 healthy individuals were used. Most people who wore spectacles took them off. They then turned to gaze at the front webcam. They were told to open and shut their eyes while staring into a camera. For 20 s, their photos were recorded.

DLA was implemented to develop the models. The base for the DLA is neural structured network. There are three layers, and they are input layer, hidden layer, and output layer of a neural network in its most basic form. These neural networks need to be trained with some training data, just like machine learning methods, before solving a specific problem. The input layer takes up the input signals and sends them to the next layer, which works based on mathematical functions and then offers the final prediction in the output layer. Information is initially sent into the input layer, which then passes on to the hidden layers. The connections between these two layers initially give weights to each input at random. After obtaining the output from the model and calculating the error, the weights are adjusted through backward propagation in order to minimize the error. Which node should be fired for feature extraction and output calculation is up to the activation function. Forward propagation is the term referring to the entire procedure. Once the output model has been obtained, and the error has been determined, weights are modified in backward propagation to lower the error. This procedure lasts for a certain number of epochs (iteration). After updating the model weights, the prediction is complete.

DLA: convolutional neural network (CNN) was used to train the eye images, and the machine learning library used was Google TensorFlow. The pictures were trained using the CNN technique. Gradient descent and static gradient descent (SGD) optimizers were the two that we employed. The following criteria were used in this investigation: the batch value = 128, learning rate = 0.001, and training data session = 1000. A group of 15,000 images are utilized for training the model, and 1000 images are supplied as test data. Experiments were carried out using an EEG dataset collected in a realistic simulation of real-world conditions. To imitate real-time data acquisition, each of the sample data taken from the pre-EEG dataset was analyzed one-at-a-time at the same sampling frequency as real-time recording. As a result, the generated artifact detection algorithm may be used without modifications for real-time artifact detection. The real dataset was captured while the study subjects worked on spot-the-difference exercises utilizing a multichannel EEG recording system with a 2048 Hz sampling rate. The vertical electrooculogram and frontal channel EEG were carefully inspected to determine the eye blinking ranges ground truth (EOG). The First 290 s of EEG data were used for the evaluation.

4.1 Convolutional neural network

Machine learning includes convolutional neural networks, sometimes known as CNN. It is a subset of the several artificial neural network models that are employed many and various purposes and datasets. A CNN is a specific type of deep learning network designed for operations like image recognition and pixel data processing. CNN is the favored network design for classifying and recognizing objects in deep learning, even though there are other types of neural networks [24]. Therefore, they are ideally suited for CV activities and can be used for applications where accurate object identification is crucial, such as face and self-driving automobile systems. Deep learning techniques rely heavily on artificial neural networks (ANNs) [24,25]. Recurrent neural networks (RNNs), which take input in sequential or time series data, are one type of ANN. It is appropriate for applications requiring speech recognition, language translation, picture captioning, and natural language processing (NLP). CNN is another neural network capable of locating significant data in both the time series data and image data. As a result, it is beneficial for mission involving images, such as pattern recognition, object classification, and image identification. To discover patterns in a picture, a CNN employs ideas from linear algebra, such as matrix multiplication. Additionally, CNN may classify audio and signal data.

The structure of CNN is comparable to the connectivity structure of the human's brain. Similar to how the human's brain has billions of neurons, CNN also have neurons, but they are structured differently in actuality, CNN's neurons are set up similarly to the brain's frontal lobe, which processes visual inputs. This configuration guarantees that the full visual field is covered, eliminating the issue with typical neural networks' piecemeal image processing that requires pictures to be given to them in low-resolution chunks CNN performs better with picture and voice or audio signal inputs than the earlier networks. Effective feature identification is a task that CNN can perform without human supervision. They are particularly good at classifying and recognizing images. Sharing weight is also another key benefit of CNNs. Convolutional neural networks also reduce computation compared to a standard neural network [25]. The TensorFlow library is used to develop the model. Abstraction is the single most important pros of TensorFlow offers for ML development. The developer can concentrate on the overarching logic of model rather than worrying about the minute details of implementing algorithms or working out how to connect the result of one function to the input of another. The details are handled in the background via TensorFlow.

4.1.1 CNN layers

The three layers that make up CNN can be thoroughly learned as convolutional, pooling, and fully connected (FC) layers. The first layer is the convolutional layer, while the final layer is the FC layer. The complication of the CNN grows from the convolutional layer to the FC layer. CNN can detect ever bigger and more intricate aspects of a picture until it successfully recognizes the entire thing as a result of the rising complexity. The convolutional layer, the central component of a CNN, is where most calculations occur. A subsequent convolutional layer may follow the first convolutional layer. A feature map or convolved feature results from the dots connected in a certain pattern. In this layer, the picture is ultimately transformed into numeric values that the CNN can understand and bring out its pertinent sequence.

4.1.2 Pooling layer

The pooling layer, similar to the convolutional layer, sweeps filter or kernel over the input data image. Contrary to convolutional layer, this layer has fewer input parameters but also causes some information to be lost. Pooling layer minimizes the dimensions of the attribute maps. Therefore, its requirements are very less with the parameters to be learned and require less network processing. The features present in a specific area are summarized in the feature map created by the feature pooling layer of a convolution layer. Therefore, subsequent actions are carried out using summarized features rather than the precisely positioned features generated by the convolution layer. Therefore, it is more resilient to differences in the placements of the features in the input data image. In a good way, this layer streamlines the CNN and boosts its efficiency. In a good way, this layer streamlines the CNN and boosts its efficiency.

4.1.3 Fully connected layer

Based on the characteristics gathered in the preceding layers, picture categorization in the CNN takes place in the FC layer. Fully linked in this context indicates that every activation unit or node of the subsequent layer is connected to every input or node from the preceding layer. CNN has yet to fully link all its layers as that would create an excessively dense network. It would cost a lot to compute, increase losses, and have an impact on output quality.

Multiple layers of a CNN are possible, and each layer trains the CNN to recognize the many aspects of an input picture. Each picture is given a filter or kernel to create an output that grows well and is more specified with one and every layer. The filters may begin as basic characteristics in the bottom levels. To examine and identify characteristics that specifically reflect the input item, the complexity of the filters increases with each additional layer. As a result, after each layer, the output of each convolved picture becomes the input data for the next layer. The CNN recognizes the picture or objects it represents in the last layer, which is an FC layer. The input data image is processed via several different filters during convolution. Each filter performs its function by turning on specific aspects of the image, after which it sends its output to the filter in the subsequent layer. The step processes are repeated for hundreds, or even thousands of layers as each layer learns to recognize various characteristics. Finally, the CNN can recognize the full object after processing all the picture data through its many layers.

A CNN, on the other side, employs parameter sharing. Every node in a CNN layer is connected to every other node. As the layers' filters traverse over the image, a CNN also has a corresponding weight, referred to as parameter sharing. Because of this, the entire CNN system requires less processing than a neural network system. A subset of machine learning called "deep learning" employs NN with a minimum of three layers. A network with numerous layers can produce more accurate findings than one with only one layer. Depending on the application, deep learning employs both RNNs and CNN. CNN is very helpful for image identification, classification, and computer vision because they produce precise results, particularly when a lot of data are involved. As the object data pass through the CNNs several layers, it also picks up the item's characteristics over time. There is no longer a requirement for manual feature extraction thanks to direct learning. CNN may be constructed on existing networks and retrained for new recognition tasks. These benefits create new possibilities for utilizing CNN in practical applications without raising computing costs or complexity. Using parameter sharing, CNNs are more computationally efficient than conventional NNs. The models may operate on any device and are simple to deploy (Fig. 10.2).

RNNs main drawback is they are unable to scale. A conventional NN could deliver acceptable outcomes for small-scale pictures with lesser color channels. As an image's size and complex growth, so does the need for processing power and resources necessary for a bigger, more extremely high-cost neural network. Additionally, overfitting becomes an issue over time when the NN attempts to extract many features from the training data. Additionally, it can wind up learning the data noise, which has an impact on how well it performs with test datasets. In the end, the NN is unable to recognize the characteristics or patterns in the dataset and, consequently, the item itself.

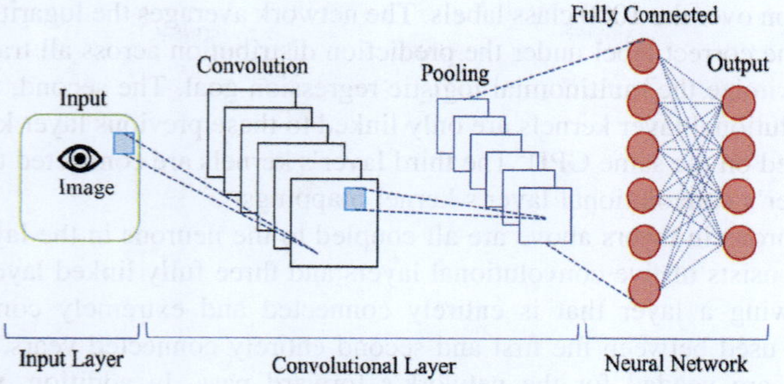

FIGURE 10.2
Architecture of CNN.

4.2 AlexNet

Eight layers make up the AlexNet, including three fully linked layers and five convolutional layers as its weights. Except for the final layer, which creates a SoftMax distribution over the 1000 class labels, ReLu activation is performed after each layer. In the first two completely linked layers, dropout is used. Max pooling is used after the first, second, and fifth convolutional layers. The next three layers from convolutional layer kernels are only connected to these kernel mappings in the preceding layer that are housed on the same graphics processing unit (GPU). The third layer's kernels are connected to all the second layer's convolutional layer's kernel mappings.

CNN views AlexNet as a turning point in image categorization. The conv and ReLu, pooling design, GPU, dropout, and parallel computing are just a few of the techniques that continue to be used in computer vision applications. The direct picture input to the classification model is AlexNet's special benefit. Edges of pictures may be automatically extracted using convolution layers, and fully linked layers can learn these features. By including additional Conv layers, it is possible to extract the crucial complexity of visual patterns effectively. In comparison to newer models like VGGNet, Google Net, and ResNet, AlexNet is NOT deep enough. Shortly after that, big convolution filters (5×5) are discouraged. The Xavier technique was eventually substituted for the normal distribution approach since it could not overcome the gradient vanishing problem. More complicated models, such as GoogleNet (6.7%) and ResNet (3.6%), perform better than it does.

The supervised learning project AlexNet produced great results. Choosing techniques that improved the network's performance, such as dropout and data augmentation, was also crucial. The AlexNet implemented ConvNets in an innovative way that is still used today, such as ReLU and dropout. Low recognition error without overfitting is difficult to achieve. There are totally eight weighted layers in the AlexNet, five of which are convolutional and three completely linked. The outcome of the final completely linked layer is sent into a 1000-way SoftMax, which creates a distribution over the 1000 class labels. The network averages the logarithmic probability of the correct label under the prediction distribution across all training samples to maximize the multinomial logistic regression goal. The second, fourth, and fifth convolutional layer kernels are only linked to these previous layer kernel mappings housed on the same GPU. The third layer's kernels are connected to all of the second layer's convolutional layer's kernel mappings.

The neurons in layers above are all coupled to the neurons in the layers below. AlexNet consists of five convolutional layers and three fully linked layers. Relu is used following a layer that is entirely connected and extremely convolutional. Dropout is used between the first and second entirely connected years. Numerous parameters are needed for the network's forward pass. In addition, we see the employment of convolution layers, which account for 95% of the calculation while only comprising 6% of the total parameters. The first significant CNN model to employ GPU training was AlexNet. This made model training quicker while compared to LeNet, AlexNet has a deeper and denser architecture (8 layers), making it more capable of extracting features (Fig. 10.3).

4.2.1 Softmax

The plot of two functions is Softmax and ReLU. The activation function acts as a "gate" in mathematics that separates the input that feeds the current neuron from its output, which travels to the next layer. Although a linear equation is straightforward to solve, it is not as powerful when it comes to solving complex issues or inferring complex functional mappings from data. Although a linear equation is straightforward, it is not as powerful when solving complex issues or inferring complex functional mappings from the data. Without the help of an activation function, a neural network is always a linear regression model. Neural networks typically employ nonlinear activation functions, which can aid in learning complex data,

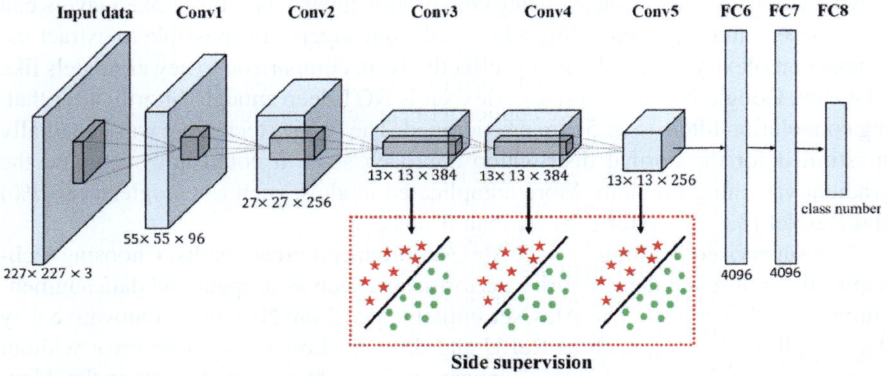

FIGURE 10.3

Architecture of AlexNet.

computing and learning nearly any function corresponding to a question, and making precise predictions. The neural network's unprocessed outputs are turned into a vector of probabilities using the SoftMax activation function, simply a probability distribution over the input classes. Think about an N-class multiclass classification task. The nonlinear activation function Softmax is likely the most straightforward of the group. Max is approximated by Softmax. It is an approximate representation of the maximum. Take note of the smooth curve that simulates the abrupt corner at 0. The range of output probabilities is SoftMax's primary benefit. The probabilities will be in the range of 0–1, and the total will be 1. In a multiclassification model, the SoftMax function provides probabilities for each class, with a high probability for the target class. The Softmax activation algorithm can handle multiple classes. It divides by the sum of the outputs for each class between 0 and 1, normalizing them. Creating a probability distribution as a result. Consequently, this provides a crystal clear likelihood that the input belongs to any given class.

The SoftMax converts input values—positive or negative or zero or higher than one—into values between 0 and 1, which can be interpreted as probabilities. Little or negative inputs are converted into small probabilities by SoftMax, and big or positive inputs into high probabilities, although they will always fall between 0 and 1. The SoftMax function is called the multiclass logistic regression or soft argmax function. This is so because the sigmoid function, used in logistic regression, has a formula in SoftMax, a version of logistic regression that can be used for multiclass classification. A classifier can utilize only the SoftMax function when the classes are mutual. The final layer of many multilayer NNs produces real-valued results that are not easily scaled and may be challenging to deal with. Because it transforms the values into a normalized probability distribution that can be shown to a user or utilized as input to other systems, SoftMax is highly helpful in this situation. It is customary to add a SoftMax function as the neural network's final layer because of this.

4.2.2 ReLU

For larger positive values of z, the slope of the ReLU function is not near zero. This hastens the convergence of the optimization. The neural network would only produce a linear input function if we applied a linear activation function or identity activation function. Because adding these levels will result in another linear function that is insufficient to model data, our neural network will behave exactly like a single-layer network, no matter how many layers it has. Additionally, our activation function ought to change. Although the slope remains 0 for negative values of z, most neurons in a neural network eventually have positive values. The AlexNet uses Rectified Linear Unit Nonlinearity, which is a key component. The standard training method for neural network models applies tanh or sigmoid activation functions. If the activation function of a neural network is nonlinear, almost any process conceivable can be represented as a functional computation. Deep CNNs might be trained significantly more quickly utilizing ReLU nonlinearity than they could be using saturating activation functions like tanh or sigmoid; these explanations are according to AlexNet. It trains faster with the help of ReLUs, which mentions ReLU function.

$$f(x) = \max(0, x) \ldots \qquad (10.1)$$

An activation function that seems and behaves like a linear function but is nonlinear is required to train deep neural networks using stochastic gradient descent with backpropagation of mistakes. Additionally, the function must be more sensitive to the activation sum input and prevent simple saturation. The answer has been floating around in the field for a while, but it was not brought to light until publication. A rectified linear activation unit, or ReLU is a node or component that implements this activation function. Networks with hidden nodes that employ the rectifier function are sometimes referred to as rectified networks. The adoption of ReLU is one of the few turning points in the deep learning revolution, such as the ways that currently make it possible to regularly create very deep neural networks. In contrast to other activation functions, it does not limit the output. This indicates that the loss of features could be better. Not the dataset itself, but the negative result of the summation of gradients is negated. As a result, since some perceptrons are inactive, the model training speed will increase further.

AlexNet deals with the high-level dataset, which may get overfitted due to the input of huge data in the processing of the model. Hence, to overcome the major issue of overfitting, we use two different methods which help to reduce the overfitting issue, and they are

1. Data augmentation,
2. Dropout.

4.2.3 Data augmentation

The training set was expanded by 2048 thanks to the authors' creation of picture translations and horizontal reflections. Additionally, they used principle component analysis (PCA) to alter the intensities of the RGB channels, which decreased the top-1 mistake rate by more than 1%. A convolutional neural network is considered invariant if it can reliably categorize objects even when positioned in various orientations and may be used in a variety of unique circumstances. A CNN can be resistant to translation, perspective, size, and lighting, for example. This is basically how data augmentation operates. We might have a dataset of images taken with few specific conditions. Our target application, however, may be used in various circumstances, such as those with changes in scale, brightness, orientation, and position. We can consider these conditions by adding more synthetically altered training data to our neural network.

Increasing the amount of pertinent data in your dataset might be beneficial. This has to do with how neural networks pick up new information. Let me give you an example to explain further. A and B eyes are the two categories in the dataset. The one represents Eye A on the left and Eye B the one on the right. Imagine the above-described dataset. Assume that every eye of A is positioned precisely. Once it is trained now that the dataset has been fed to the neural network, and hoping for outstanding results. It lowers the price of data acquisition. The price of labeling data is decreased. It raises the model's forecast precision. It avoids a data shortage. Better data models are framed. It lessens the overfitting of data. It gives data models variety and adaptability. It improves the data models' capacity for generalization. It aids in fixing the classification's class imbalance problem.

4.2.4 Dropout

A dropout was AlexNet's second strategy for preventing overfitting. With a probability of 0.5, the output of each hidden nodes is set to zero. The neurons that have "fallen out" in this way do not engage in either the forward pass or the backpropagation. As a result, each time an input data is provided, the neural network samples a different architecture. With this method, neurons are turned off with a certain probability. This implies that the neurons that have been "shut off" do not contribute to the forward pass or the backpropagation throughout each iteration. The key advantage of this approach is that it stops all neurons in a layer from improving their weights simultaneously. This random adaptation keeps the neurons from convergent toward the same objective, decorrelating the weights. The activations of the hidden units become sparse, which is a second property revealed for the use of dropout and is likewise a desirable trait.

The Dropout layer serves as a mask, keeping all other neurons viable but removing some neurons' contributions to the following layer. If we apply a Dropout layer to the input vector, some of its characteristics are eliminated; however, if we apply it to a hidden layer, some hidden neurons are eliminated. Because they avoid overfitting the training data, dropout layers are crucial in the training of CNNs. If they are absent, the first set of training samples has an excessively large impact on learning. As a result, traits that only show in subsequent samples or batches would not be learned. Therefore, CNN would not discover the existence of straight lines.

4.3 Computer vision

Computers can now extract meaningful information from digital images, movies, and other visual inputs thanks to AI domains of computer vision. With this information, computers may then act or offer suggestions. Like artificial intelligence offers machines the ability to think, computer vision enables machines to sense, observe, and comprehend. As human eyesight has been around longer than computer vision, it has the edge over the latter. The benefit of human sight is that it can acquire the skills necessary to distinguish between things, determine their distance from the viewer, determine whether they are moving, and determine whether an image is accurate throughout the course of a lifetime. Computers can learn to accomplish equivalent tasks far more quickly thanks to computer vision than they could with retinas, optic nerves, and a visual brain. As a system that has been trained to check goods or keep track of a production asset may assess hundreds of items or procedures per minute while finding hidden flaws or errors, it can soon outperform people. A lot of data are needed for computer vision. It runs data analysis repeatedly until it recognizes photos and distinguishes between objects. For instance, many tire pictures and tire-related information must be provided to teach a computer to recognize automobile tires. This is especially true of flawless tires.

Using machine learning and algorithmic models, a computer may be taught how to comprehend the context of visual data. If enough data are fed through the trained model, the computer will read the data and train itself to discern between distinct images. The use of algorithms enables a computer to learn on its own rather than requiring programming to recognize a picture. A CNN facilitates the "seeing" capabilities of a machine learning or deep learning model by breaking down images into pixels values with labels or tags. Using the labels, convolutions are run, and predictions are made based on what the system is "seeing." The neural network performs convolutions and continuously assesses the precision of its predictions up until they begin to come true. The next step is to identify or see images in a manner comparable to how people do. A CNN first detects crisp shapes and fundamental forms before adding details as it iteratively verifies its predictions, similar to how a person might see a picture from a distance. A CNN helps to comprehend individual images. RNNs are utilized in video applications in a manner similar to this to help computers understand the relationships between the visuals in a sequence of frames. Society is made easier by computer vision systems' rapid completion of boring and repetitive activities. Very well-trained computer vision systems will never make a mistake. Faster delivery of high-quality goods and services will follow from this. Computer vision will eliminate the need for mediocre goods and services, saving businesses the expense of rectifying their problematic processes. The human aspect of exhaustion is not present in computers and cameras; thus, it is not an issue with them. The effectiveness is typically constant and unaffected by other variables like illness or emotive state. The same computer system can be seen in a variety of settings and tasks. Additionally, it can be seen among many other alternatives, warehouse tracking and supply delivery in manufacturers, as well as scanned photographs in the medical sector. The process of computer imaging takes less time and has a lower mistake rate. It lowers the expense of hiring and training specialized personnel to perform the tasks that computers will perform at the rate of hundreds of workers.

4.4 Image recognition

The technique of identifying and analyzing photos to identify objects, places, people, or other items observable in one's natural environment is known as image recognition, a subset of computer vision. The main objective is to see the items similarly to how a human brain would. These elements are what image recognition aims to find and assess before making judgments based on the results. Contrarily, the term "computer vision" is wider and includes all methods for gathering, evaluating, and interpreting data from the real world for use by machines. Similar to how people do it, image recognition analyses each pixel in an image to extract pertinent information. A broad variety of objects may be detected and recognized by AI cameras using computer vision training. Humans use a natural neural network to recognize pictures, which helps individuals detect objects in photos based on prior experiences. The artificial neural network also helps robots recognize images. An AI neural network's several layers of cells can interact with one another. The kind of information required also affects the structure and architecture of a neural network. Because it involves deep learning, neural networks, and sophisticated picture recognition algorithms to be practical for machines, image recognition is more difficult than you may think. The effort and intervention needed from human agents can be greatly reduced as a result. The amount of time required to complete particular tasks, such as identity verification or signature validation, is significantly decreased by an automated system. With the help of image recognition technologies, you may complete more tasks in a shorter time and reduce other costs, such as manpower, in the process. Additionally, real-time visual data analysis gives business owners insightful information that enables them to act quickly on information gleaned through image recognition technologies.

We need to first train algorithms that can learn and forecast with high accuracy before computer image recognition is even viable. Think about how you can tell an image of a cat is a cat just by looking at it, yet an image recognition system works differently. The confidence score is a measurement of a machine's ability to perceive something as being 75% kitten, 10% puppy, and 5% like other comparable styles as an animal according to shared traits. Additionally, to correctly forecast the item, the computer must first comprehend what it sees before analyzing it and making a final prediction by comparing it to previous training. Machines can see and analyze picture material in ways that humans are not. Machines perceive pictures as either a vector or raster, which is a collection of pixels, in contrast to humans. This objective is achieved for machines that can precisely characterize what is happening in pictures with the help of convolutional neural networks. To recognize images, AI is now supported by convolutional neural networks. However, the issue of how to make various pictures recognizable to AI emerges. The reason is that these images have been classified using the proper data labeling methods to provide high-quality training datasets.

5. Experimental results

The TensorFlow DLA was able to identify between the open and closed conditions of test eye pictures; after training with 15,000 photos, we acquired a probability of 90.1%–90.7%. The gradient Descent optimizer scored 90.05, the SGD optimizer scored 88.01, and the Adam optimizer scored 90.7 for each TensorFlow optimizer, as shown in Table 10.3. When there was no caretaker present, the patient looked at the camera while closing their left eye, which caused a sound to be produced to summon the caretaker. There was a 100% success rate. In this model, we implemented

Table 10.3 Result table.

Training data	Validation data	Model used	Class	Accuracy
12,927	1433	CNN	Close_look, forward_look, left_look, right_look	88.4%
14,360	2870	Alex-net	Close_look, forward_look, left_look, right_look	90%

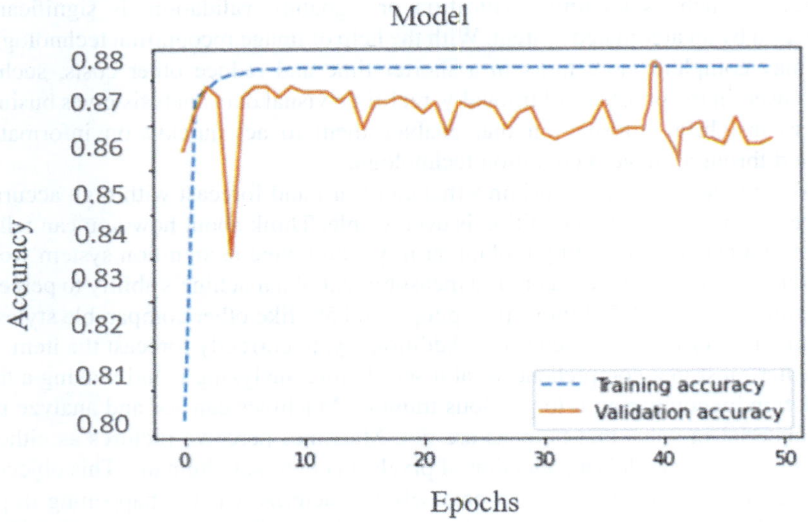

FIGURE 10.4

Accuracy diagram from CNN.

two kinds of algorithms Convolutional neural network and AlexNet. The validation accuracy of CNN in Fig. 10.4 explains that epoch 5 had the least accuracy, and it improved to 88% at epoch 39, as shown in Table 10.4.

The validation accuracy of CNN in Fig. 10.5 explains that epoch 6 had the least accuracy, and it got improved to 90% at epoch 14. In CNN, we used only two layers for the model; hence, the result had only minimum accuracy when compared to AlexNet. AlexNet has eight layers in it, and the accuracy of the model has increased to its extent. AlexNet always selects the related parameters for the model. AlexNet does not utilize the then-standard SoftMax function; instead, it employs rectified linear units (ReLU). ReLU has an advantage over SoftMax in training time; a CNN using ReLU was able to achieve a 25% error six times more quickly. To prevent overfitting, dropout is widely utilized in massive linear transformations.

Table 10.4 Accuracy table.

Number of epoch	1	2	3	4	5	6	7	8	9	10	11	12
CNN	50.31	62.37	72.68	80.21	81.6	83.3	83.2	84.2	84.8	85	86	87
Alex-net	78.3	82.2	83.2	84.6	85.7	86.2	87.8	88.1	88.6	88.9	89.6	90.4

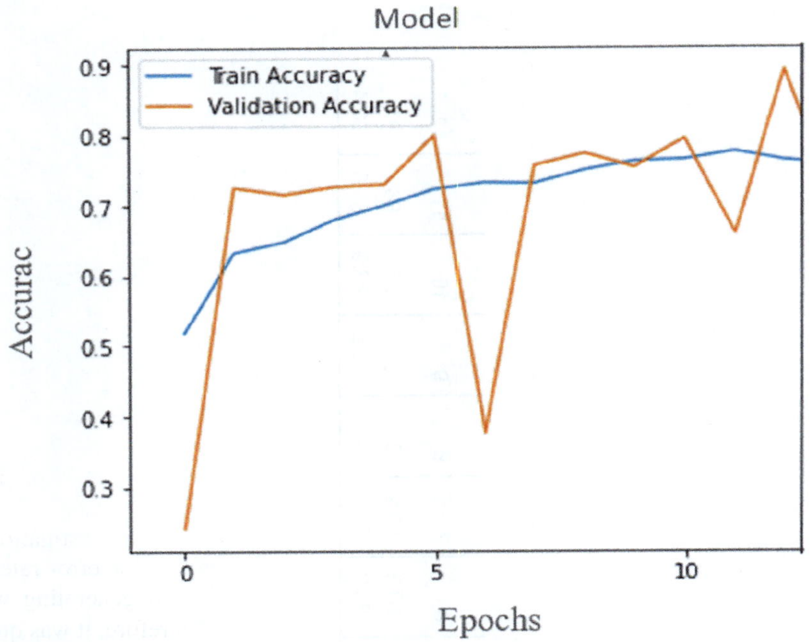

FIGURE 10.5

Accuracy diagram of AlexNet.

6. Discussion

Both image-based and nonimage-based techniques are used to determine if an eye is closed. EEG is a typical nonimage-based method. This method has the benefit of quick data collection. To get around these issues, several studies have been carried out utilizing an image-based methodology. A video-based approach and a single-image-based method were the two categories of image-based procedures to assess eye-opening status. Several video-based techniques have shown excellent accuracy results. They took a long time and required a lot of calculation, though. Single-image methods were separated into training and nontraining approaches. Iris detection was one of the most widely used single-image-based approaches and a nontraining technique. The key benefit of these methods was that they required more training. The process was not necessary. The disadvantage of these approaches was that if a suitable extraction fell short of producing the necessary details required to identify the eye state from a picture, the improvement of picture acquisition while gaining needed characteristics would be limited. However, compared to video-based systems, single-image methods and training techniques had a processing time advantage and were more accurate than nontraining methods.

Finding the best feature extraction methodology required using training techniques. We thought about using CNN-based techniques to overcome this issue. In the area of image identification, CNN is a form of deep learning that is often employed. CNN made it easier for filters and classifiers to automatically acquire the best features from training data. The extraction was learned automatically from the training set of data. In various CNN model types, we employed TensorFlow. Google created the open-source deep learning platform TensorFlow. In the industrial setting, it is frequently employed for both research and business purposes. Although it is not perfect, our DLA pretty reliably recognizes whether the eye is open or closed. CNN's approach for verifying ocular status has a stated inaccuracy rate. In our investigation, eye photos were taken in a single second, and it was chosen whether or not to open the eyes for 1 s if three or more similar images were obtained. When three or more images are identified, a distress signal in the form of noise is emitted after the left eye has remained closed for a duration of 3 s. A noise was emitted to signal for help after the left eye had been closed for 3 s. The identifying eye status was terminated by pressing the "log end" button. The single-second eye status is inaccurately identified.

The single-second eye's condition is incorrectly described. The investigation's eye state detection error rate per second was 0.002% due to the 1.3% error rate of eye status detection from an image. The success rate of sound generating was 99.99% if the operation was carried out continuously for 3 s. Therefore, it was quite likely that one eye was shut constantly for 3 s. Building instruments just to see if the eyes opened once or the eyes closed once or twice would be pointless. It is crucial to ensure uninterrupted ocular conditions in particular places, nevertheless. Monitoring the eye state, for instance, may be necessary for a gadget that detects tiredness while driving or other hazardous tasks. The condition of your eyes may have an impact on the outcome of your examination in the medical field. EEG is a typical example. When an EEG test is conducted, waves might change based on the condition of the eye. Therefore, before AI was generally recognized, a handful of tests were done on a system for verifying the eye condition during the EEG. Even if the patient did not have a piece of particular monitoring equipment, using our methodology, the eye state could be ascertained by the time via the webcam. In severely disabled individuals who can only move their eyes, it is imperative to ascertain whether the eyes would be open state or closed. These conditions include locked-in syndrome, amyotrophic lateral sclerosis, and complete cervical spinal cord injury. These patients need constant care.

As caretakers are people, they cannot keep an eye on the patients at all times. Consequently, a gadget that produces sound to call the caretaker, when necessary, benefits both the patient and the caretaker. As a result, we created this application. The program in our trial merely generated a sound to summon the caretaker. Recently, quadriplegic patients have benefited from the development and use of commercial ETCSs. These patients can receive assistance from ETCSs in several additional ways in addition to phoning the caretaker. However, due to the limited number of patients required, such hardware and software are costly. Although the technology employed in this study cannot do eye-tracking, it can be utilized with a camera on a typical office laptop. This system is constrained in various ways. The widely used mobile operating systems Android and iOS for the iPhone are initially incompatible with this program, which is first exclusively deployed on Windows. Other software cannot be used in parallel with this application. Second, the program cannot detect eyeballs on the face if the user moves their head too much while it is operating. Additionally, the program might not be able to distinguish the face if the webcam is too close to or too far away from the face. Additionally, facial and eye recognition are impacted by lighting conditions. The incidence of patients with these illnesses is often relatively low; hence the number of patients who used this method was modest. Thus, more research involving more patients is required.

7. Conclusion

Our AI-based eye state detection program is incredibly precise and produces unbiased evaluations. For paraplegic patients who find it difficult to call a caretaker, we designed a calling system utilizing AI and got good results. The approach might have significant applications in other clinical practice and research fields. Consequently, further research utilizing AI is required to increase patient convenience. After training with the 15,000 photographs we gathered, the TensorFlow DLA was able to distinguish between the closed and open state of eye pictures for testing with a probability of 90.1%–90.7%. The patient stared at the camera while shutting their left eye, which resulted in a sound being created to call for the caretaker when there was no one there. Convolutional neural networks and AlexNet are two types of algorithms that we used in this model.

The system was developed to provide aid to people who are fully paralyzed and help them when there is a need to call a caregiver. Patients with quadriplegia typically use their voices to summon help. However, patients with severe quadriplegia are tracheostomized and unable to speak. To contact caretakers, some patients need different communication means. The brain–computer interface recognizes the exact eye movement for 3 s, and it fails to establish the exact response when the eyes or half open and when a person tries to sleep. We can provide other useful methods which may improve the usability of the software, which may help the patients.

References

[1] Chang W-D, Lim J-H, Chang-Hwan IM. An unsupervised eye blink artifact detection method for real-time electroencephalogram processing. Physiological Measurement 2016;37(3):401.

[2] Wezel J, Anders G, Webb AG, van Osch MJP, Beenakker JWM. Automated eye blink detection and correction method for clinical MR eye imaging. Magnetic Resonance in Medicine 2017;78(1):165–71.

[3] Lee JH, Kang T, Choi BK, Han IH, Kim BC, Ro JH. Application of deep learning systems into the development of communication devices for the quadriplegic patient. Korean Journal of Nutrition 2019;15(2):88.

[4] Mouthami K, Nirmala Devi K, Bhaskaran VM. Sentiment analysis and classification based on textual reviews. In: 2013 international conference on information communication and embedded systems (ICICLES). IEEE; 2013. p. 271–6.

[5] Jenifa G, Indu S, Jeevitha C, Kiruthika V. Visitor/intruder monitoring system using machine learning. AI and IoT-Based Intelligent Automation in Robotics 2021:295–304.

[6] Kim KW, Hong HG, Nam GP, Park KR. A study of deep CNN-based classification of open and closed eyes using a visible light camera sensor. Sensors 2017;17:E1534.

[7] Esteva A, Kuprel B, Novoa RA, Ko J, Swetter SM, Blau HM, et al. Dermatologist-level classification of skin cancer with deep neural networks. Nature 2017;542(7639):115–8.

[8] Feng R, Marcus B, Mocco J, Oermann EK. Deep learning guided stroke management: a review of clinical applications. Journal of Neurointerventional Surgery 2018;10(4): 358–62.

[9] Krizhevsky A, Sutskever I, Hinton GE. Imagenet classification with deep convolutional neural networks. Communications of the ACM 2017;60(6):84–90.

[10] Chang W-D, Cha H-S, Kim K, Chang-Hwan I. Detection of eye blink artifacts from single prefrontal channel electroencephalogram. Computer Methods and Programs in Biomedicine 2016;124:19–30.

[11] Pult H, Riede-Pult BH, Murphy PJ. A new perspective on spontaneous blinks. Ophthalmology 2013;120(5):1086–91.

[12] Lugo Z, Pellas F, Blandin V, Laureys S, Gosseries O. Assessment of needs, psychological impact and quality of life in families of patients with locked-in syndrome. Brain Injury 2017;31(12):1590–6.

[13] Li Y, Ma Z, Lu W, Li Y. Automatic removal of the eye blink artifact from EEG using an ICA-based template matching approach. Physiological Measurement 2006;27(4):425.

[14] Klein A, Skrandies W. A reliable statistical method to detect eyeblink-artefacts from electroencephalogram data only. Brain Topography 2013;26(4):558–68.

[15] Galley N, Schleicher R, Galley L. Blink parameters as indicators of driver's sleepiness–possibilities and limitations. Vision in Vehicles 2004;10:189–96.

[16] Tafreshi M, Fotouhi AM. A fast and accurate algorithm for eye opening or closing detection based on local maximum vertical derivative pattern. Turkish Journal of Electrical Engineering and Computer Sciences 2016;24(6):5124–34.

[17] Van Middendorp JJ, Watkins F, Park C, Landymore H. Eye-tracking computer systems for inpatients with tetraplegia: findings from a feasibility study. Spinal Cord 2015;53(3): 221–5.

[18] Verleger R. Valid identification of blink artefacts: are they larger than 50 μV in EEG records? Electroencephalography and Clinical Neurophysiology 1993;87(6):354–63.

[19] Veiga JJD, O'Reilly M, Whelan D, Caulfield B, Ward TE. Feature-free activity classification of inertial sensor data with machine vision techniques: method, development, and evaluation. JMIR mHealth and uHealth 2017;5(8):e7521.

[20] Fukuda K, Stern JA, Brown TB, Russo MB. Cognition, blinks, eye-movements, and pupillary movements during performance of a running memory task. Aviation Space & Environmental Medicine 2005;76(7):C75–85.

[21] Choi J-S, Bang JW, Heo H, Park KR. Evaluation of fear using nonintrusive measurement of multimodal sensors. Sensors 2015;15(7):17507–33.

[22] Dataset. https://www.kaggle.com/datasets/kayvanshah/eye-dataset.

[23] Mathur M, Jindal V, Wadhwa G. Detection of the multi-scale images using SRP network-A review. In: 2021 third international conference on inventive research in computing applications (ICIRCA). IEEE; 2021. p. 1578–83.
[24] Nakayama Y, Shimizu T, Mochizuki Y, Hayashi K, Matsuda C, Nagao M, et al. Predictors of impaired communication in amyotrophic lateral sclerosis patients with tracheostomy-invasive ventilation. Amyotrophic Lateral Sclerosis and Frontotemporal Degeneration 2016;17(1–2):38–46.
[25] Olczak J, Fahlberg N, Maki A, Sharif Razavian A, Anthony J, Stark A, et al. Artificial intelligence for analyzing orthopedic trauma radiographs: deep learning algorithms—are they on par with humans for diagnosing fractures? Acta Orthopaedica 2017; 88(6):581–6.

CHAPTER

11

Motor imaginary tasks-based EEG signals classification using continuous wavelet transform and LSTM network

A. Ananthi[1,4], M.S.P. Subathra[1,4], S. Thomas George[2,4], N.J. Sairamya[3,5], J. Prasanna[2,4], and P. Manimegalai[2,4]

[1]*Department of Robotics Engineering;* [2]*Department of Biomedical Engineering;* [3]*Department of Psychology;* [4]*Karunya Institute of Technology and Sciences, Coimbatore, Tamil Nadu, India;* [5]*University of Calgary, Calgary, Canada*

1. Introduction

Brain–computer interface (BCI) has helped to bridge humans and computers because they enable deliberate computer control through studying brain signal patterns. EEG is a signal resource in BCI for recognition for its minimal cost, fine time resolution, and ease of use [1–3]. Although BCIs could be utilized in a wide range of ways to aid control, motor imagery (MI) BCIs are of particular research importance. This is due to the high potential of its application in industries including gaming, neuroprosthetics, and rehabilitation that require translation of the person's thinking and visualized actions [1]. The captured EEG data of right and left motor imagery are used to control a cursor on a computer display. It could serve as a unique communication channel to substitute a user's impaired motor control. As a result, it may help to increase the value of life for diseased persons. The feature extraction aids in the identification of EEG signal biomarkers regarding that imagination. This is significant because EEG data contain a combination of noises. Investigators have done with a variety of feature extraction methods, such as wavelet transform, entropy, phase spatial filter, autoregression, and so on [3]. The motor imaginary-based brain–computer interface system reads EEG data and then finds real-time responses to EEG signals. When analyzing EEG data, researchers discover nonstationarity, noise, artifacts, and a multichannel recording paradigm. Signal processing gets slower as the number of signals to analyze increases. Investigators are now working on motor imaginary EEG-based brain–computer interfaces due to their extensive scope [4–6]. The MI approach gathers signals from volunteers' ideas

Computational Intelligence and Deep Learning Methods for Neuro-rehabilitation Applications
https://doi.org/10.1016/B978-0-443-13772-3.00013-3

while doing a task, such as a limb movement. When picturing the movement of an autonomous limb, a range of stimulations occurs in a particular cerebral area. Volume conduction converts these changes into electrical impulses that travel through numerous brain tissues [7]. The use of MI is to regulate the movement of the right and left fists, the right and left feet, and a few other movements including the opening and closing of the hand. The aforementioned movements exhibit distinct variations in specific regions and the extent of cortical activation, suggesting they possess significant and unique characteristics [8,9].

Banghua et al. [10] proposed a technique using wavelet packet decomposition and common spatial pattern to extract features. The authors attained classification accurateness of 80% and 92% for a database of authors-prepared data and the BCI competition 2008 database, respectively. Yang et al. applied Fisher wavelet packet decomposition (F-WPD) and common spatial pattern to every subject individually. When compared to nonsubject-based techniques, the accuracy is increased by about 14% [11]. Few research works are dealing with the problem that employs the single hidden layer feedforward network (SLFN) to display velocity in several activities such as brain activity detection [12,13], image division [13], and facet categorization [14].

Hettiarachchi et al. [15] retrieved the subband coefficient and subband energy average of WPD as preliminary features and compared the results using the Fisher distance criterion (FDC). The researchers experienced their suggested technique on BCI Competition 2003 Database I b and attained a classification accuracy of 59.4% using the feature selection technique.

The augmented complex-valued common spatial pattern (CCSP) system was proposed to be responsible for common difficult impulses using noncircular probability distributions, and an accuracy of 72.37% was achieved for 56 out of 109 subjects with CCSP well found with robust uncorrelated transform classification [16]. Fast Fourier transformation (FFT) was used to extract six statistical features and feed into k nearest neighbor (KNN) technique to classify the signal [5]. Magnitude squared coherence (MSC) is thought of as a superior feature for evaluating the relation between impulses generated in various brain areas, resulting in diverse patterns. The KNN algorithm were used to achieve 100% accuracy in biometric identification [17]. Utilizing the CSP method for feature extraction and SVM for the classifier, an accuracy rate of 81% has been procured with 35 top overall subjects. When 16 or 20 channels were used instead of all 64 channels, classification accuracy improved [18]. Researchers commonly employed wavelet transform (WT) for fine feature examination, and SVM was applied to categorize the features [19]. The channels were minimized to 48 out of 64, and with random forest, the accuracy achieved with the fewest electrodes was 90%. The genetic algorithm suggested selecting a feature, with 34 individuals in addition to 19 channels extracted using power spectral density also classified using SVM and KNN [20]. Another research the accuracy attained was 85.5% based on the new classifier LightGBM with minimum channels [21].

Deep learning (DL) is a system that efforts to compete with the performance of the humanoid brain by creating a compound [22] connected to NN and then improving a standard model that could grip a wide range of inputs. DL attempts to recoup from the faults of ordinary NN by overlapping the entire training information. The gained EEG information immediately fed into the DNN. The goal of deep learning is to feed raw data into DNNs with little or no preparation [6]. DL differs from machine learning in that it chains feature extraction, and selection, followed by classification into a single process. Deep neural network (DNN) training involves a substantial number of parameters and hyperparameters, which can result in longer training durations compared to other techniques. Additionally, it requires significant hardware resources to accommodate these computational demands [9,23]. The issue is handled by employing innovative computing strategies, for instance graphical processing units (GPU).

Researchers have just recently begun to modernize the use of deep learning algorithms in the analysis of huge datasets [18,23]. DL techniques, as opposed to classic NNs, can learn a large number of needed parameters to be more strongly on familiar terms with vital and detailed information. Deep learning is used in the study [7] with CNN and conventional fully connected layers for learning the features and as a classifier respectively to achieve determined accuracy of 86.49%, 79.25%, and 68.51% of 2, 3, and 4 classes, respectively. With 32 electrodes, the CNN-GRU classification system reached 99.90% to 100% means CRR is also 99.17% with five channels [24]. While the combined CNN (EEGNet, Shallow ConvNet Deep ConvNet) classifier had better accurateness by 84.1 and 83.8% while verified on 103 participants, the computational time was four times than previous approaches [25]. The chess board transformation (CBT) method outperformed an azimuthal projection method and SVM when extracting spatial and frequency characteristics using DCNN and LSTM. The CBT had an accuracy of 68.72%, while the azimuthal projection and SVM had an accuracy of 68.13% and 64.64%, respectively [26]. DCNN and squeeze-and-excitation blocks were utilized to develop a deep EEG impulses descriptor to calculate the effect of the motor mission on the EEG impulse during biometric verification, yielding a 0.1% of EER rate [27]. Researchers [16] employed CNN approaches in two, three, and four classrooms and reached maximum accuracy of 82.88% in two classes, 74.75% in three classes, and 65.82% in four classes. The author presents a unique MI classification strategy followed by CWT and CNN in this research. By combining time frequency and electrode location, a CWT with three mother wavelets is utilized to create an informative EEG picture. Then, CNN is built to identify MI tasks while also reducing computation complexity [28]. The author has reviewed all the papers related to PhysioNet dataset that has been used in deep learning and machine learning techniques [29]. The researchers [30–37] have given their results based on various parameters using statistical features, entropy, and FD features for classification signals.

Among the main factors contributing to long-term impairment is stroke [38,39]. The majority of motor deficits brought about by stroke could be recovered relative to deficiencies brought on by other neurological conditions or severe brain injuries. Furthermore, two BCI techniques have been demonstrated to aid in rehabilitation for people whose muscle function has also been completely lost or severely damaged due to sickness or trauma [40]. By facilitating connections with environments using brain signals instead of muscles, BCI devices can compensate for the absence of regular neuromuscular impulses. In Austria, a motorized hand device built on synchronized BCI and driven by continuing EEG function has been used [41]. However, BCI innovation is still advanced and has just lately been used in research experiments. BCI techniques have been proven to restore motor control; they may also aid in restoring brain plasticity by influencing motor learning [42]. The much more reliable action of central nervous system plasticity [43–45] after brain damage.

To date, no research had been conducted to investigate features deconstructed with CWT. As a result, in our study, we proposed: Deep learning models are used in a novel way to classify the EEG output. With a variety of features, it can achieve higher classification accuracy. The suggested method begins with feature extraction from EEG signals using CWT. In our work, we employ the LSTM classification technique. The suggested CWT-based log energy entropy as a single feature for classifying EEG data for left and right fists performs significantly better. In this research, we present a CWT approach and an LSTM for (MI-BCI) EEG data classification and discussed how the proposed work is useful for the neurorehabilitation of stroke patients.

2. Material and methodology

2.1 Dataset (PhysioNet)

The EEG database from (Refs. [46,47]; https://archive.physionet.org/physiobank/database/eegmmidb/) 64 channel contains 1500 signals (each sampled at 160 samples per second) from EEG recordings utilizing 10–10 electrode placement systems, which were performed by 109 patients. Each participant used their imagination to move their left or right fist when an object appeared on the left or right side of the display, respectively. The trials were where there was no object and were designated as relaxation. The target was closed as well opened with the imagination until it vanished from the display. Every trial continued for 4 s, succeeded by a brief period of inactivity. Each user provided an average of 150 EEG trials, with about equal scattering of the right, left, or rest labels. As mentioned early, the location of the 64 electrodes was shown in Fig. 11.1.

(a) Task 1: As soon as an object is made known in the display, the participant opens and closes the suitable paw centered on either an object's face on the left or right side, till the object is wiped out. When a mission is completed, the participant can wind down.

(b) Task 2: As soon as an object is made known in the display, the participant pictures open and close the correct paw centered on either the object faces on the left or right side, till the object is wiped out. When a mission is completed, the participant can wind down.

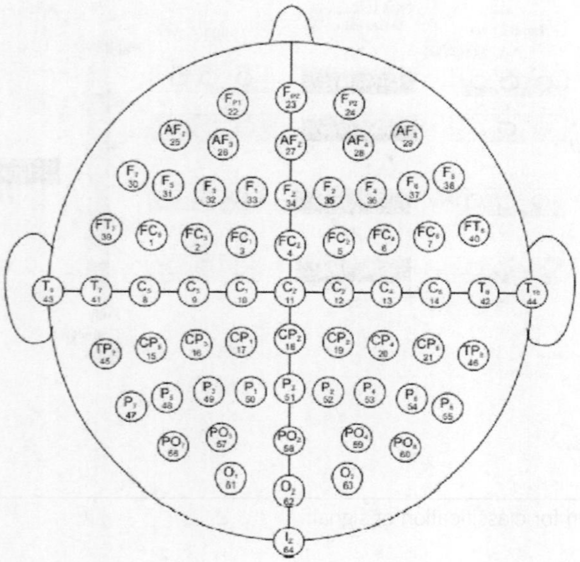

FIGURE 11.1
Location of EEG electrodes according to the 10–20 system.

(c) Task 3: As soon as an object is made known in the display, the volunteer opens and closes both hands if the object faces on the upper side, opens and closes two feet if the object faces toward the lower till the object is wiped out. When a mission is completed, the participant can wind down.
(d) Task 4: As soon as an object is made known in the display, the target pictures open and close both paws if the object faces on the top, opening and closing two feet if the object faces toward the lower till the object is wiped out. When a mission is completed, the participant can wind down.

The proposed method is to categorize the signals to identify the imagining of the volunteer as well as label the trials into one of the two labels: right and left classes. The preprocessed signal is fed into CWT to extract 14 features including statistical features, Fractal dimension, Entropy features, and so on. And those features have been given to the classifier (LSTM model) integrating all extracted features. The previous technique is to utilize the database to train and validate a particular model. On the other hand, we show that 92.4% classification accuracy be accomplished with our proposed technique.

2.2 EEG preprocessing and feature extraction

The EEG signals are picked up from the human brain. The raw human brain signals have noise and so the redundant noises are eliminated, and then the features of

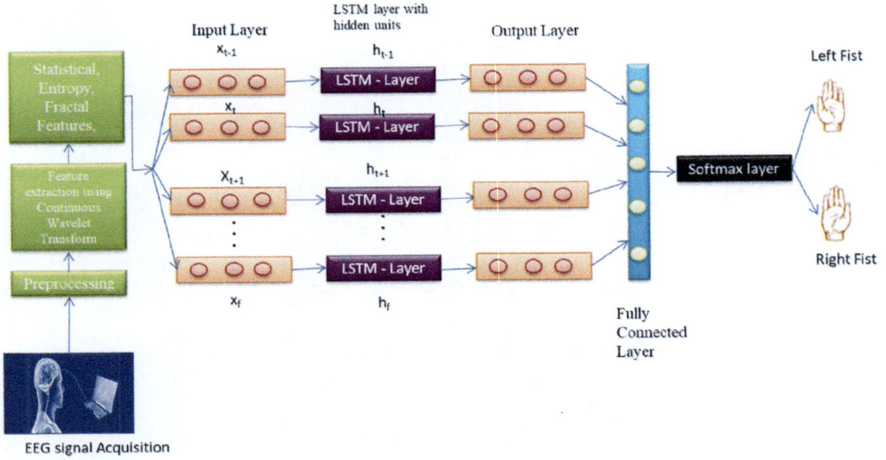

FIGURE 11.2

Workflow diagram for classification of signal.

signals are mined by CWT, subsequently which the two-class information's classified using LSTM. Fig. 11.2 shows the Workflow diagram for the Classification of the Signal.

2.2.1 Continuous wavelet transform

A wavelet study has brought together a windowing method with variable-sized areas. WD, as shown in the given diagram, familiarizes the idea of scale as a potential substitute to maps and frequency of signal data in a time-scale plane (Fig. 11.3). It is the same as the time–frequency (t–f) plane utilized in short term Fourier transform. In the time–frequency domain, individual time-scale agrees to a specific frequency

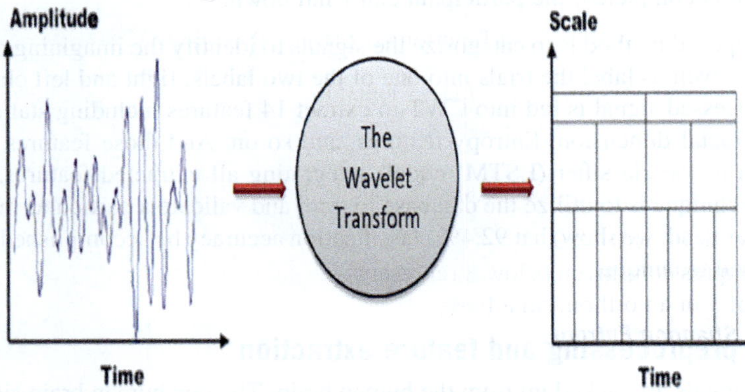

FIGURE 11.3

Concept of time-scale plane.

range. A wavelet is a short-duration waveform. In contrast to sine waves, they prolong from − to + infinity, the wavelets are confined waves that last for a set amount of time.

Arithmetically, the CWT function *f*(*T*) is stated as the integral transform of *f*(*T*) with a family of wavelet functions, ψA, *B*(*T*):

$$\mathrm{CWT(A,B)} = \frac{1}{\sqrt{a}} \int_{-\infty}^{+\infty} f(T) * \phi\left(\frac{T-B}{A}\right) dT \tag{11.1}$$

$$A \in R^{+} - \{0\}, B \in R$$

CWT is also stated with shifted versions of the wavelet function (ψ) as shown:

$$\mathrm{CWT(scale, position)} = \int_{-\infty}^{+\infty} f(t) * \phi(scale, position, t)dt \tag{11.2}$$

The function (*T*) is known as the mother wavelet, The family of functions *a*, *b* (*t*) is known as the daughter wavelets.

The daughter wavelet is produced by scaling and moving the mother wavelet. The scale factor a represents the function's scaling (*t*), whereas the shift factor *b* represents the function's temporal translation. In matrix C, the CWT generates several specific wavelet coefficients. It is critical to understand the importance of determining the scale parameter of CWT.

2.2.2 Features

Entropy is the most widely used model in various fields, mostly in the field of signal processing. In this work, we are using 14 features including statistical feature, fractal dimensions, entropy features, and so on. All the features are discussed below.

2.2.2.1 Entropy features

Log energy entropy (7) here make the most of the feature vector in the classification stage. Entropy is mainly employed in signal processing techniques as it can extract the required information.

Entropy *E* is an additive cost function so $E(0) = 0, E(s) = \sum_i E(s_i)$

$$E(s_i) = \log(s_i^2) \tag{11.3}$$

So,

$$\mathrm{E(s)} = \sum_i \log(s_i^2) \tag{11.4}$$

With convention log(0) = 0.

In the expressions given below, *s* represents the signal and $(s_i)_i$ represents the coefficients of *s* in an orthonormal basis.

2.2.2.1.1 Shannon Entropy

$$E(s_i) = s_i^2 \log(s_i^2) \tag{11.5}$$

So,

$$E(s) = \sum_i s_i^2 \log(s_i^2) \tag{11.6}$$

With convention 0log(0) = 0.

2.2.2.1.2 Sure entropy

$$E(s) = n - \#\{\text{i such that } |s_i| \leq p\} + \sum_i (s_i^2, p^2) \tag{11.7}$$

$0 \leq p$, where p is the threshold value

Here "s_i" implies signal coefficient in an orthonormal basis. The value of entropy is more than 1, and the module has a capacity to expose additional data regarding the signal and is required to decompose further, so that could attain a simple frequency module of signal (Shinde, 2004).

Entropy usage is to obtain unique features of the signal and minimize the size of the feature vector compared with node coefficients.

2.2.2.2 Statistical features

2.2.2.2.1 *Mean* Mean is defined in theory as the number of items divided by the total number of elements in a collection. Calculating the mean provides us with a good understanding of the entire set of data. The Mean can be written as

$$\mu_j = \frac{1}{n}\sum_{j=1}^{n} k_j \tag{11.8}$$

The parameter k_j stands for the arithmetic score that occupies location j in the density function, and n stands for the number of accessible codes.

2.2.2.2.2 *Standard deviation* The standard deviation measures how much a set of data can vary or be dispersed. In contrast to a high standard deviation, which indicates a wider range of values, a low standard deviation indicates that results tend to be close to the collection's mean, also known as the expected value.

$$\sigma_i = \sqrt{\frac{1}{n}\sum_{i=1}^{n}(k_i - \mu_i)^2} \tag{11.9}$$

The parameter k_i stands for the arithmetic score that occupies location i in the density function, and N stands for the number of accessible codes.

2.2.2.2.3 *Skewness* Skewness is a degree of uniformity, or more precisely, the lack of uniformity. A distribution or dataset is said to be uniform if it appears identical to the left and right of the center point. If the curve is shifted to the left or right, it is said to be skewed. Skewness can be expressed as

$$S_j = \frac{\dfrac{1}{n\sum_{j=1}^{n}\left(k(j) - \mu_j\right)^2}}{n\sigma^3} \tag{11.10}$$

The parameter μ_j stands for the arithmetic score that occupies location j in the density function, and n stands for the number of accessible codes.

2.2.2.2.4 Normalized standard deviation The normalized standard deviation (NSD) is obtained by simply dividing the standard deviation by the mean (or coefficient of variation). The standard deviation is expressed as a percentage of the mean. This statistic allows for a more accurate comparison of the spread of a variable with a huge mean and correspondingly huge SD and other variable with a reduced mean and correspondingly reduced SD. The NSD is written as

$$n\sigma_i = \frac{\sigma_i}{i - i} \tag{11.11}$$

2.2.2.2.5 Kurtosis Kurtosis indicates whether the data reached its maximum or minimum in relation to a normal distribution. In other words, kurtosis indicates whether or not high values are present in the tails of a distribution. Kurtosis can be expressed as

$$kr_j = \frac{\dfrac{1}{n \sum_{j=1}^{n} \left(k(j) - \mu_j\right)^4}}{n\sigma^4} \tag{11.12}$$

The parameter j stands for the arithmetic score that occupies location j in the density function, and n stands for the number of accessible codes.

2.2.2.2.6 Energy The energy of EEG signals is well defined with the square of the EEG signal magnitude, the wrapping of squared EEG signal magnitude, or the integral of squared EEG signal magnitude. Energy can be represented as

$$e_j = \frac{1}{m} \sum_{j=1}^{m} |k_j|^2 \tag{11.13}$$

The parameter kj stands for the arithmetic score that occupies location j in the density function, and m stands for the number of accessible codes.

2.2.2.2.7 Normalized energy Normalized energy is defined as the current energy e_i divided by the peak of the current tracker value (k_i) where ne_i is the normalized value of the energy at the i-th frame.

$$ne_i = \frac{e_i}{length(k_i)} \tag{11.14}$$

2.2.2.3 Fractal dimension

2.2.2.3.1 Higuchi fractal dimension Higuchi is to evaluate the fractal dimension (FD), here d taken as the input of quantity made a discrete time series. Eq. (11.4) is well defined as

$$y_L^N = X(N), X(N+L), X(N+2L), X(N+3L)\ldots, X\left(N + \left[\frac{n-N}{L}\right]L\right) \tag{11.15}$$

here $N = 1, 2, 3, \ldots, L$.

$\left[\frac{n-N}{L}\right]$ refers to Gauss notation, L and N refer to integers.

Length of y_L^N of each time is in Eq. (11.15):

$$l_N(L) = \frac{1}{L}\left(\sum_{i=1}^{\left[\frac{n-N}{L}\right]} |y(N+iL) - y(N+(i-1)L)| \frac{n-1}{\left[\frac{n-N}{L}\right]L} / L\right) \quad (11.16)$$

here term (Eq. 11.16) denotes

$$\frac{n-1}{\left[\frac{n-N}{L}\right]L} \quad (11.17)$$

denotes a normalization factor in the length of FD. $l(K)$ denotes mean value over L sets of $l_N(L)$ as demonstrated below Eq. (11.18) [19]:

$$l(L) = \sum_{N=1}^{i} l_N(L) \quad (11.18)$$

With all the above equations, fractal dimension d of Higuchi is evaluated as a slope for linear regression of plots [7].

2.2.2.3.2 Katz fractal dimension Katz FD resolves an issue using the yardstick general unit-making technique. The variable is used to represent the average space between continuous points [8]. It is calculated with Euclidean distances (ED) between continuous points [48]. As a result, the equation is inscribed as Eq. (11.19) as given below:

$$d = \frac{\log_{10}\left(\frac{l}{A}\right)}{\log_{10}\left(\frac{D}{A}\right)} = \frac{\log_{10}(N)}{\log_{10}\left(\frac{D}{l}\right) + \log_{10}(N)} \quad (11.19)$$

Here A is the mean space among continuous points, and $N = 1$, separated by the number of continuous points in the waveforms [24].

2.3 Classification

2.3.1 Long short-term memory technique

RNNs face problems in learning enduring dependencies. The long short-term memory (LSTM) model is similar to RNNs in that it can solve the disappearing gradient problem clearly. LSTM models effectively expand RNN memory, allowing them to retain and learn enduring input dependencies. Memory extension is capable to retain knowledge for an extended duration, allowing for writing, reading as well as erasing data from memories. Long short-term memory is also known as a "gated" cell. Here gate represents the capability to choose either to have or discard information in memory. The LSTM models pull together essential features beginning with inputs as well as store the data over a longer duration. The decision to delete or have the data is based on the weight values given to the material throughout the teaching phase. An LSTM technique will now learn which information to keep and which to discard.

There are three gates in the LSTM model. The first one is named as forget gate, the second is the input gate, and finally, the gate is named as output gate. The first one s_i forget gate that removes the current data, the input gate insists on the scope of new information being added in memory, and the output gate controls whether the current data in the cell adds to the output. Each gate in the process is explained in detail and the Fig. 11.4 shows pictorial representation of LSTM.

1. **Forget gate:** As previously stated, the sigmoid function is used here to remove information from the LSTM's memory. The values of g_{u-1} and y_u are used to make this decision. The values of e_u are in the middle of 0 and 1, here 0 denotes entirely discarding the well-read value and 1 indicates stabilizing all values. The result is calculated as follows:

$$\mathrm{e_u} = \sigma\left(X_{e_g}[\mathrm{g_{u-1}}], X_{e_y}[y_u], B_e\right) \tag{11.20}$$

   $\mathrm{B_e}$ value is constant and so it is known as bias value.
2. **Input gate:** The input gate is utilized to enforce the scope of new information to be added to the memory. It contains two layers: one is the sigmoid layer and the other is "tanh" layer. Here sigmoid layer selects the data to be modernized, and tanh layer generates a vector of novel data to be additionally given to the LSTM memory. The following are the results of two layers:

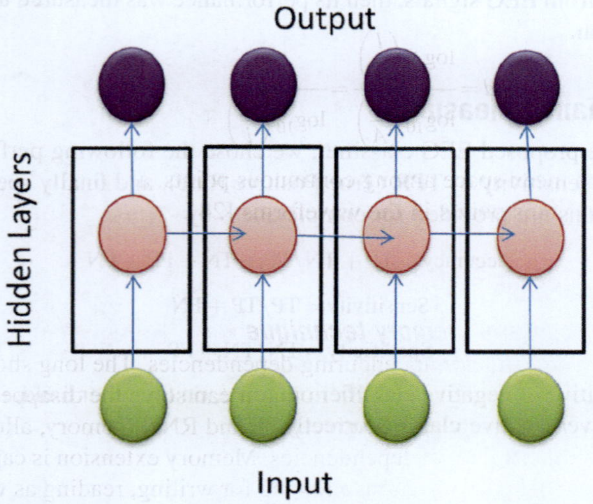

FIGURE 11.4
Architecture of LSTM.

$$j_u = \sigma\left(X_{j_g}[g_{u-1}], X_{j_y}[y_u], B_j\right) \tag{11.21}$$

$$d_u = \tanh\left(X_{d_g}[g_{u-1}], X_{d_y}[y_u], B_d\right) \tag{11.22}$$

where j_u indicates either the value essential to update or not, and d_u implies a vector of fresh data that are given in the memory of LSTM. Grouping of two layers is to update LSTM memory. Here the present information is unable to remember by utilizing layers of forget gate. This is expressed as an equation as shown below.

$$d_u = e_u * d_{u-1} + j_u * d_u \tag{11.23}$$

here e_u is the output of forget gate, where the values are in the middle of 0 and 1.

3. **Output gate.** The output gate employs the sigmoid layer to determine which values of LSTM memory are routed to this gate. After that, a nonlinear tanh function is used to plan the data from 1 to 1. Lastly, the outcome is multiplied with the sigmoid layer outcome. The end result is evaluated with the equation as given below:

$$a_u = \sigma\left(X_{a_g}[g_{u-1}], X_{a_y}[y_u], B_a\right) \tag{11.24}$$

$$g_u = a_u * tanh(d_u) \tag{11.25}$$

where a_u denotes output value, and g_u implies value from -1 to 1.

2.3.1.1 Results and discussions

Section 3 describes how the CWT-LSTM algorithm was applied in MATLAB R2018. Here, 40 EEG signals are investigated to examine the recommended classifier's accuracy in classifying from the respective signals. The classifier was fed with CWT features from EEG signals, then its performance was measured using a 10-fold computer vision.

2.4 Performance measures

To evaluate the proposed EEG classifier, we chose the following performance metrics: accuracy, sensitivity or TPR (True Positive Rate), and finally specificity. Their formal definitions are provided for completeness by

$$\text{Accuracy} = \text{TP} + \text{TN}/\text{TP} + \text{TN} + \text{FP} + \text{FN}$$

$$\text{Sensitivity} = \text{TP}/\text{TP} + \text{FN}$$

$$\text{Specificity} = \text{TN}/\text{TN} + \text{FP}$$

A false positive or negative classifier outcome, on the other hand, is one that predicts the positive/negative class incorrectly.

2.5 Classification performance evaluation

The EEG motor movement Imaginary database, which included signals from both the left and right fists, was split into two parts: a train and a test set. The EEG data were preprocessed, and CWT features were extracted and fed into the best-performing classifiers to assign class labels. Following that, the systematic labels corresponding to every EEG signal were fed into LSTM, which ultimately determined the class label. A 10-fold Cross Validation was used to estimate classification performance in terms of Accuracy, Sensitivity, and Specificity as shown in Table 11.1, we achieved promising classification performance for EEG datasets using the proposed scheme. Using CWT, the 14 features were extracted from the EEG signals. Following feature extraction, we feed all of the values into LSTM. Now, we start with a simple network with a single hidden layer as well as one neuron, progressively increasing the number of neurons as well as hidden layers until we achieve better classification performance. Hidden units of 50 achieve the highest classification accuracy. In this task, use an input size of one and connected the sequence input layer to an LSTM layer with 50 hidden layers. The result of the LSTM layer is passed to the fully connected layer, which is then accompanied by the softmax layer. At last, the outcome is given to the classification layer to improve accuracy. To achieve the best results, we have used the Adam optimizer with the tanh activation function and the highest 50 epochs.

For reference, a comparison with other previously worked results is provided. Accuracy is a major competitor in performance measurement. As a result, we tried to compare the classification accuracy to related work as in the Table 11.2. The proposed CWT work is preceded by feature extraction and then classification to LSTM. The features used here outperform with 92.4% accuracy. Furthermore, using LSTM

Table 11.1 Performance of the proposed methodology.

Method	Features extracted	Features selected	Accuracy	Specificity	Sensitivity
CWT + LSTM	Statistical, entropy, and fractal dimension features	23,394	92.4	90.1	89.5
CWT + LSTM	Statistical features	13,368	89.7	86.3	81.6
CWT + LSTM	Entropy features	5013	85.9	83.5	80.1
CWT + LSTM	FD features	3342	78.6	77.2	75.3

Statistical features: Min, Mean, Max, standard deviation, normalized SD, energy, normalized energy, skewness, and kurtosis.
Entropy features: ILog energy entropy, sure entropy, and Shannon entropyFractal dimension features: Higuchi and Katz.

Table 11.2 Comparison with previous work.

Title	Methods	Accuracy
Dose et al. (2018)[7]	CNN	80.32%
Li et al. (2019)[23]	DCNN-mVGG	96.48%
Fan et al. (2020)[16]	QNET	82.88%
Roots et al. (2020)[25]	CNN	83.8%
Li et al. (2020)[23]	CNN-GRU	97.36%
Hou et al. (2019)[13]	Scout ESI and CNN	94.51%
Abenna et al. (2022)[21]	Delta-CSP-LGBM	98.17%
Our work	CWT + LSTM	92.4%

as a classification method results in improved accuracy. Eventually, our work in variable spaces of CWT level, number of CWT hidden layers, as well as the number of neurons to achieve highest accuracy function as an exact method and pays to achieve in our technique.

The primary objective of the suggested method would be to deploy motor imaginary-based BCI devices for such neurorehabilitation of stroke patients. As a result of this strategy, a BCI system was used, which converted the patient's motor intention into a control signal to run a device to aid stroke patients. The EEG signal was elicited by the patient's intent. The goal was to use the user's EEG data to control a man's walking direction. The machine could be controlled by the user's fictitious signals. Images on the left and right hands symbolized turning to the left and right, respectively. By offering feedback to assist rehabilitation, enhanced conventional poststroke recovery may take place in real-time settings. The user could be trained to balance the device's operation before and after. The primary work of designing the neurorehabilitation device is to classify the signals using machine and deep learning algorithms.

3. Conclusion

We present a methodology for BCI-based EEG analysis based on the continuous wavelet transform. The technique's performance is validated using the PhysioNet dataset. An LSTM completes the classification, and its accuracy is proportional to the amount of work done. To achieve maximum classification accuracy, the technique's factors such as the number of neurons and the number of hidden layers are taken into account in this work. In the database, the proposed method achieves 92.4% accuracy with 14 features including statistical, fractal dimension, and entropy features. The final result outperforms all other results available on the same and different databases. Overall, the suggested methodology of using CWT and LSTM to classify the signal is an effective tool for BCI-based EEG analysis. A forthcoming research goal would develop a real-time intelligent system for medical use. The proposed system is designing the neurorehabilitation device is to classify the EEG signals using artificial techniques for stroke persons.

References

[1] Padfield N, Zabalza J, Zhao H, Masero V, Ren J. EEG-based brain-computer interfaces using motor-imagery: techniques and challenges. Sensors 2019;19(6):1423.

[2] Pfurtscheller G, Neuper C, Schlogl A, Lugger K. Separability of EEG signals recorded during right and left motor imagery using adaptive autoregressive parameters. IEEE Transactions on Rehabilitation Engineering 1998;6(3):316−25.

[3] Nandish M, Stafford M, Kumar PH, Ahmed F. Feature extraction and classification of EEG signal using neural network based techniques. International Journal of Engineering and Innovative Technology (IJEIT) 2012;2(4):1−5.

[4] Park C, Cheong Took C, Mandic DP. Augmented complex common spatial patterns for classification of noncircular EEG from motor imagery tasks. IEEE Transactions on Neural Systems and Rehabilitation Engineering 2013;22(1):1−10.

[5] Alpturk EK, Kutlu Y. Analysis of relation between motor activity and imaginary EEG records. arXiv; 2021. preprint arXiv:2101.10215.

[6] Alazrai R, Abuhijleh M, Alwanni H, Daoud MI. A deep learning framework for decoding motor imagery tasks of the same hand using EEG signals. IEEE Access 2019;7:109612−27. https://doi.org/10.1109/access.2019.2934018.

[7] Dose H, et al. An end-to-end deep learning approach to MI-EEG signal classification for BCIs. Expert Systems with Applications 2018;114:532−42.

[8] Dubey JD, Arora D, Khanna P. Reducing electrodes based on decision tree classification for EEG motor movement data. 2018.

[9] Mammone N, Ieracitano C, Morabito FC. A deep CNN approach to decode motor preparation of upper limbs from time−frequency maps of EEG signals at source level. Neural Networks 2020;124:357−72. https://doi.org/10.1016/j. neunet.2020.01.027.

[10] Bang-hua YANG, Wen-yu LU, Mei-yan HE, Li L. Novel feature extraction method for BCI based on WPD and CSP. Chinese Journal of Scientific Instrument 2012;33(11):2560−5.

[11] Yang B, Li H, Wang Q, Zhang Y. Subject-based feature extraction by using Fisher WPD-CSP in brain−computer interfaces. Computer Methods and Programs in Biomedicine 2016;129:21−8.

[12] Hu J, Xiao D, Mu Z. Application of energy entropy in motor imagery EEG classification. International Journal of Digital Content Technology and its Applications 2009;3(2):83−90.

[13] Shinde A, Hou Z. A wavelet packet based sifting process and its application for structural health monitoring. Structural Health Monitoring 2005;4(2):153−70.

[14] Vidal JJ. Real-time detection of brain events in EEG. Proceedings of the IEEE 1977;65(5):633−41.

[15] Mohammed AA, Minhas R, Wu QJ, Sid-Ahmed MA. Human face recognition based on multidimensional PCA and extreme learning machine. Pattern Recognition 2011;44(10−11):2588−97.

[16] Fan C-C, et al. Bilinear neural network with 3-D attention for brain decoding of motor imagery movements from the human EEG. Cognitive Neurodynamics 2021;15(1): 181−9.
[17] Singh B, Mishra S, Tiwary US. EEG based biometric identification with reduced number of channels. In: 17th international conference on advanced communication technology (ICACT). IEEE; 2015. 2015.
[18] Handiru VS, Prasad VA. Optimized bi-objective EEG channel selection and cross-subject generalization with brain−computer interfaces. IEEE Transactions on Human-Machine Systems 2016;46(6):777−86.
[19] Shah RG, Rutu N. Hand movement classification using motor imagery eeg. International Journal of Innovative Research in Science, Engineering and Technology 2016; 5(6).
[20] Saibene A, Gasparini F. GA for feature selection of EEG heterogeneous data. arXiv preprint arXiv:2103.07117; 2021.
[21] Abenna S, Nahid M, Bajit A. Motor imagery based brain-computer interface: improving the EEG classification using Delta rhythm and LightGBM algorithm. Biomedical Signal Processing and Control 2022;71:103102.
[22] Olcay BO. Murat Özgören, and Bilge Karaçalı. "On the characterization of cognitive tasks using activity-specific short-lived synchronization between electroencephalography channels. Neural Networks 2021;143:452−74.
[23] Zhu X, Li P, Li C, Yao D, Zhang R, Xu P. Separated channel convolutional neural network to realize the training free motor imagery BCI systems. Biomedical Signal Processing and Control 2019;49:396−403.
[24] Wilaiprasitporn T, et al. Affective EEG-based person identification using the deep learning approach. IEEE Transactions on Cognitive and Developmental Systems 2019;12(3):486−96.
[25] Roots K, Muhammad Y, Muhammad N. Fusion convolutional neural network for cross-subject EEG motor imagery classification. Computers 2020;9(3):72.
[26] Fadel W, et al. Chessboard EEG images classification for BCI systems using deep neural network. In: International conference on bio-inspired information and communication technologies. Cham: Springer; 2020.
[27] Mota MRF, et al. A deep descriptor for cross-tasking EEG-based recognition. PeerJ Computer Science 2021;7:e549.
[28] Lee HK, Choi Y-S. Application of continuous wavelet transform and convolutional neural network in decoding motor imagery brain-computer interface. Entropy 2019;21(12): 1199.
[29] Ananthi A, et al. A review on-EEG signals by motor imagery based brain computer interface. In: AIP conference proceedings. 2670. AIP Publishing LLC; 2022.
[30] Priya SJ, et al. Local pattern transformation based feature extraction for recognition of Parkinson's disease based on gait signals. Diagnostics 2021;11(8):1395.
[31] Catherine Joy R, et al. Detection of adhd from eeg signals using different entropy measures and ann. Clinical EEG and Neuroscience 2022;53(1):12−23.
[32] Sairamya NJ, Subathra MSP, Thomas George S. Automatic identification of schizophrenia using EEG signals based on discrete wavelet transform and RLNDiP technique with ANN. Expert Systems with Applications 2022;192:116230.
[33] Sairamya NJ, et al. A new approach for automatic detection of focal EEG signals using wavelet packet decomposition and quad binary pattern method. Biomedical Signal Processing and Control 2021;63:102096.

[34] George ST, et al. Classification of epileptic EEG signals using PSO based artificial neural network and tunable-Q wavelet transform. Biocybernetics and Biomedical Engineering 2020;40(2):709–28.

[35] Subathra MSP, et al. Detection of focal and non-focal electroencephalogram signals using fast Walsh-Hadamard transform and artificial neural network. Sensors 2020;20(17):4952.

[36] Sairamya NJ, et al. Performance evaluation of discrete wavelet transform, and wavelet packet decomposition for automated focal and generalized epileptic seizure detection. IETE Journal of Research 2021;67(6):778–98.

[37] Jose S, et al. Robust classification of intramuscular EMG signals to aid the diagnosis of neuromuscular disorders. IEEE Open Journal of Engineering in Medicine and Biology 2020;1:235–42.

[38] Sims NR, Muyderman H. Mitochondria, oxidative metabolism and cell death in stroke. Biochimica et Biophysica Acta 2010;1802(1):80–91.

[39] Donnan GA, Fisher M, Macleod M, et al. Stroke. Lancet 2008;371(9624):1612–23.

[40] Janis JD, Jonathan RW. Brain-computer interfaces in neurological rehabilitation. The Lancet Neurology 2008;7(11):1032–43.

[41] Pfurtscheller G, Guger C, Müller G, et al. Brain oscillations control hand orthosis in a tetraplegic. Neuroscience Letters 2000;292(3):211–4.

[42] Kübler A, Birbaumer N. Brain-computer interfaces and communication in paralysis: extinction of goal directed thinking in completely paralysed patients? Clinical Neurophysiology 2008;119(11):2658–66.

[43] Nudo RJ. Mechanisms for recovery of motor function following cortical damage. Current Opinion in Neurobiology 2006;16(6):638–44.

[44] Wolpaw JR, Tennissen AM. Activity-dependent spinal cord plasticity in health and disease. Annual Review of Neuroscience 2001;24:807–43.

[45] Ziemann U, Ilic TV, Pauli C, et al. Learning modifies subsequent induction of long-term potentiation-like and long-term depression-like plasticity in human motor cortex. Journal of Neuroscience 2004;24(7):1666–72.

[46] Schalk G, McFarland DJ, Hinterberger T, Birbaumer N, Wolpaw JR. BCI2000: a general-purpose brain-computer interface (BCI) system. IEEE Transactions on Biomedical Engineering 2004;51(6):1034–43.

[47] Goldberger AL, Amaral LA, Glass L, Hausdorff JM, Ivanov PC, Mark RG, Stanley HE. PhysioBank, PhysioToolkit, and PhysioNet: components of a new research resource for complex physiologic signals. Circulation 2000;101(23):e215–20.

[48] Fraschini M, et al. Robustness of functional connectivity metrics for EEG-based personal identification over task-induced intra-class and inter-class variations. Pattern Recognition Letters 2019;125:49–54.

Further reading

[1] Huang GB, Zhou H, Ding X, Zhang R. Extreme learning machine for regression and multiclass classification. IEEE Transactions on Systems, Man, and Cybernetics, Part B (Cybernetics) 2011;42(2):513–29.

[2] Huang GB, Wang DH, Lan Y. Extreme learning machines: a survey. International Journal of Machine Learning and Cybernetics 2011;2(2):107–22.

[3] Pan C, Park DS, Yang Y, Yoo HM. Leukocyte image segmentation by visual attention and extreme learning machine. Neural Computing & Applications 2012;21(6):1217–27.
[4] Sweldens W. The lifting scheme: a custom-design construction of biorthogonal wavelets. Applied and Computational Harmonic Analysis 1996;3(2):186–200.
[5] Hettiarachchi IT, Nguyen TT, Nahavandi S. Motor imagery data classification for BCI application using wavelet packet feature extraction. In: International conference on neural information processing. Cham: Springer; November 2014. p. 519–26.
[6] Kayikcioglu T, Aydemir O. A polynomial fitting and k-NN based approach for improving classification of motor imagery BCI data. Pattern Recognition Letters 2010;31(11): 1207–15.
[7] Yen GG, Lin KC. Wavelet packet feature extraction for vibration monitoring. IEEE Transactions on Industrial Electronics 2000;47(3):650–67.
[8] Birbaumer N, Hinterberger T, Kubler A, Neumann N. The thought-translation device (TTD): neurobehavioral mechanisms and clinical outcome. IEEE Transactions on Neural Systems and Rehabilitation Engineering 2003;11(2):120–3.
[9] Scherer R, Lee F, Schlogl A, Leeb R, Bischof H, Pfurtscheller G. Toward self-paced brain–computer communication: navigation through virtual worlds. IEEE Transactions on Biomedical Engineering 2008;55(2):675–82.
[10] Hochreiter S, Schmidhuber J. Long short-term memory. Neural Computation 1997;9(8): 1735–80.
[11] Graves A, Schmidhuber J. Framewise phoneme classification with bidirectional LSTM and other neural network architectures. Neural Networks 2005;18(5–6):602–10.
[12] Ieracitano C, et al. A hybrid-domain deep learning-based BCI for discriminating hand motion planning from EEG sources. International Journal of Neural Systems 2021; 31(09):2150038.

CHAPTER

12

Enhancing human brain activity through a systematic study conducted using graph theory and probability concepts on a hydar prehistoric organism

C. Alvino Rock, A. Hepzibah Christinal and D. Abraham Chandy
Karunya Institute of Technology and Sciences, Coimbatore, Tamil Nadu, India

1. Introduction

Here eating disorders (EDs) are extreme psychological conditions having cognitive, psychopathological, and neurobiological support [1,2]. In recent years, more works are made to explain the neurobiological relations for the diseases through both functional and structural neuroimaging methodologies. As briefly shown in previous testimonies [3,4], neuroimaging study in eating disorders points out that their neural relationships better showed as changes in spatial distributions and joined neural machines other than malfunctions in one and spatially far away places. This proof is constant with the broad observation that structural and functional brain variations in psychological disorders can be better justified on the basis of covariance models other than localization [5]. A strong instrument to define the arrangement of brain networks basis on their covariance models in recent times was given by connectomic methodologies, which checks out the brain as a complicated network and its changes in the properties pointing out global or regional architecture [6]. One of the major uses of using connectomic instruments is to check brain structure and function that they permit proving brain location as not as discrete and solitary elements but as things that speak to everybody based on the topological behavior of the connections.

The mathematical topic used mostly in connectomics to depict the topological behavior of brain locations is graph theory, a separate area of maths that proves the rules and the relationship available between nodes and the edges getting connected [7]. Assume brain locations as nodes and their connections functional or structural as edges, graph theory explains their topological policies and rules

Computational Intelligence and Deep Learning Methods for Neuro-rehabilitation Applications
https://doi.org/10.1016/B978-0-443-13772-3.00010-8

pointing out various categories, which are separated according to their integration, separation, or middle behaviors. A proper equilibrium is maintained between the integration and separation rules of a network ensuring to combine well its wiring cost with communication between topologically far away locations [8]. The business between integration and segregation behaviors of a network topology is described shortly by the Small-World Index (SWI), which shows the role that a network held between in the continuum intermediate regular and irregular machine topologies [9]. Another main characteristic of the brain network topology is shown by nodes that are heavily interconnected within the network, thus providing its universal structure and function, and that are mainly shown as network hubs [10]. More interconnectivity of hub locations contributes to their heavy metabolic cost, which is heavier than their peripheric locations. However, the central point of hubs is also proven to calculate a stronger problem of brain damage because dysfunctions in any brain location are most likely to be widespread through more connected node locations. Therefore, likely brain disorders show higher metabolic affects like diseases in eating which result in hubs assigning of hubs. The separation, integration, and centrality behavior of a network are not constantly the same during the development phase and change constantly respective to the needs imposed by the various maturation stages and by the continuous allotment of the highest level of cognitive methodologies. Also, their equilibrium varies frequently because of the beginning and continuation of brain disorders, which affect the structures that require more functional cost, well known as hubs. Eating disorders are of specific behavior from this point of view which bloom during adolescence period, which are depicted by vast metabolic results [11].

For such problems, any modifications done in the topological behavior of the brain for these disorders must be notedly proved, which is described both by modifications that happen during neurodevelopmental trajectories stages and as resulting disorder continuation. The use of multimodal imaging methodologies can help in separating this complex situation because modifications are done in different structural or functional indices that hold various stability during the development stages and a varying sensation to environmental problems [12]. Also, longitudinal check shows the examining of patients at various stages of the disease (i.e., during acute stages or later relieving from the hospital), as well as the evaluation of graph rules alterations changes high and rapid weight loss, can be pointedly used for knowing how the policies that manage the network architecture differs by time and in various stages of the disease [13]. Parallelly the possibility of calculating the topological behavior of the brain network, graph theory tools make us evaluate the power of connections between nodes in case-control variations, via a network-based-statistics connectome methodology [14]. Therefore, the methodology does not show the separation or integration behavior of a network but it points out showing intermediate-group comparisons, the availability of modified connections in subnetworks of two or more region locations. Wholesome connectomics instruments have been vastly used in various psychological and neurological disorders and in eating diseases as well, letting for strengthening their neurobiological base. In the current scenario, we aimed at pointing out current innovations shown in MRI-based connectomic concepts of eating disorders, which use graph theory instruments.

2. Methodology

The methodology part describes "active" concept in "active inference," beginning with why building the action in the framework which is one of the important concepts in artificial neural networks [15]. Next, we will look at how the action is build, leading to code so we can test it for ourselves. We have also discussed how to compare active inference to other machine learning frameworks. Many research papers are published on the free energy principle which quickens the knowledge and research on active inference and the free energy principle.

2.1 Bayesian inference

Bayesian inference to the rescue: the brain uses Bayesian inferences on the world using Bayes' theoremto recalculate the probability for a hypothesis x as more proof or sensory observations y are gained. We recalculate our beliefs on the world based on the proofs that we observe. The free energy principle assumes the Bayesian brain hypothesis where probability play a main role. Also, the brain needs to respond with not trained information in the real world.

2.2 Probability density

For the brain to cope with not trained information, it is not enough to work with just a single-point estimate for x. The brain has to encrypt all values of x but in probability theory this is recorded with a probability distribution: A probability distribution $\mathrm{p}(x)$ $\mathrm{p}(x)$ is a list of results and their matching probabilities of xx. Probability distributions are of two types. The probability mass function is used in the discrete areas (e.g., set of possible results is discrete denoting a coin toss or a dice roll. The probability density function is used in continuous areas where possible results can be taken on numbers. Continuous variables as used in the free energy papers and also used the terms "probability density" or simply "density."

We would explain it with an instance like the distance xx a child throws varies with time. If we put all the distances in ascending order, we notice a pattern where few distances are more probable/common than others. If we make the list elaborate we get continuous numbers and when we convert this list to a graph where x-axis is the distance thrown and the y-axis the probability we obtain as probability density, depicted as $\mathrm{p}(x)\mathrm{p}(x)$. These patterns we get can have different shapes but a similar pattern is visualized like a bell curve (like in a clock tower) and it is called a Gaussian or normal probability density. Normally, it is the most similar probability density observed in day-to-day life, hence denoted by the name normal. For instance: average height of a species of trees, income of people in a group of

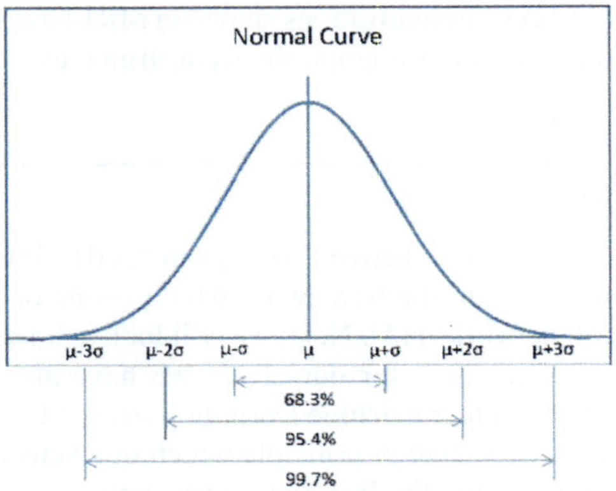

FIGURE 12.1

Normal distribution.

population, mean grade of students, etc. The free energy principle or active inference takes Gaussian probability densities called the Laplace approximation or normal distribution, which makes mathematics simple (Fig. 12.1).

The normal probability density is mathematically described with just two parameters using the following equation:

$$N(\mu, \sigma^2) = \frac{1}{\sqrt{2\pi\sigma^2}} e^{-\frac{(x-\mu)^2}{2\sigma^2}}$$

Mean μμ—the mean value of all points, for instance the average distance the child throws the ball. Standard deviation σσ—the points deviate from the mean, for instance if a child began learning to throw the ball have more standard deviation and when trained more the throws will be more consistent and hence a lower standard deviation with long distance and high mean. The deviations noted as: variance—simply the squared standard deviation: σ^2 Precision—inverse of the variance: $\frac{1}{\sigma^2}$. A higher precision denotes lower variance with a high spike in the bell curve. So instead of remembering all possible values for a variable, the brain uses the μ and σ to show the probabilities of proofs on the world. For instance, thanks to the distribution principles, 68% of the information lies within 1σ from the mean, 95% within 2σ from the average, and 99.7% within 3σ from the average. This allows instant estimating chances, a part that the brain does to save time and effort. Average μ and control signal u have different meanings.

3. Hydar influence in brain activity

All the flow diagram shows prediction, prediction error, and prediction error minimization for the active inference algorithm which helps us visualize active inference for one active inference capsule (Fig. 12.2) [16].

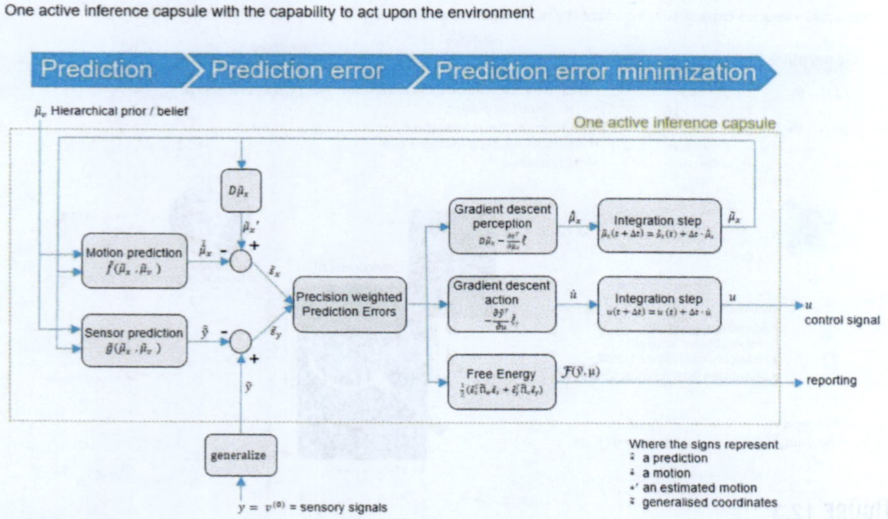

FIGURE 12.2

Active inference flow diagram.

In traditional control theory for instance Kalman estimates the hidden state (x) to determine the control signal (u) by comparing it to the measured state. In active inference, however, the hidden state estimate μ_x is a mix of the hidden state estimate and desired state μ_v. Because the motion μ_x by the function of motion (f) in the generative model is both expressed in terms of (μ_x) and (μ_v), it is denoting the motion to the operational specification denoted by the hierarchical prior μ_v. Also, the perception portion and active portion unfold parallelly, doing both free energy minimization parallel. In this first example of action in active inference, Hydar beliefs are encrypted in the biased generative model where it moves toward its hierarchical prior (μ_v). Hydar's propulsion gives the vertical speed ((Fig. 12.3) ($u = x'$).

As Hydar is a basic prehistoric imaginative aquatic ancestor, which we have with the most simple generative models to see if Hydar can still live which goes back to the evolutionary argument. Despite complex generative processes in the environment, Hydar still takes action to maintain the temperature in a specific bandwidth with a simple generative model.

3.1 Movement of actual robot

When an actual robot moving using the active inference code gets inserted as a node in a robotic operating system environment, it points out a Jackal robot with skid-type driving similar to a tank, which turns by moving the right or left wheels faster. The hidden is a vector holding the velocities of the x, y directions and the rotation velocity. The sensory observation is a vector holding left and right wheel speeds with the rotation velocity originating from the IMU (Fig. 12.4).

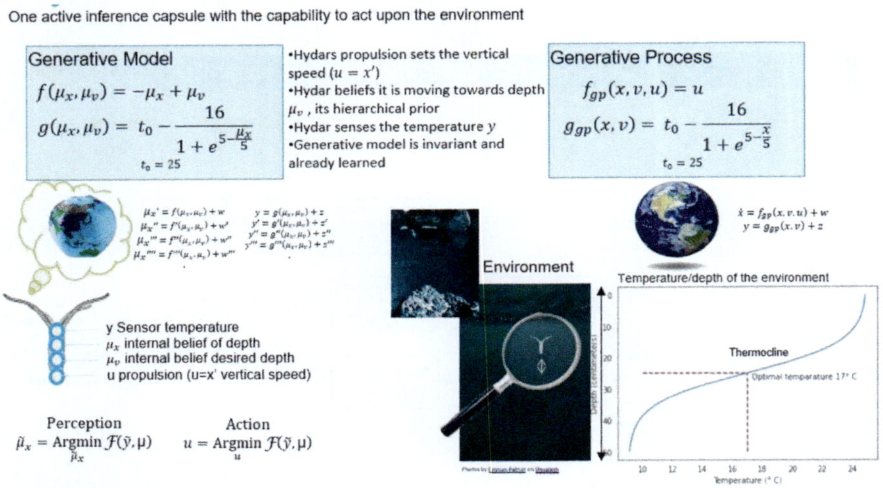

FIGURE 12.3

Hydar acts on the environment.

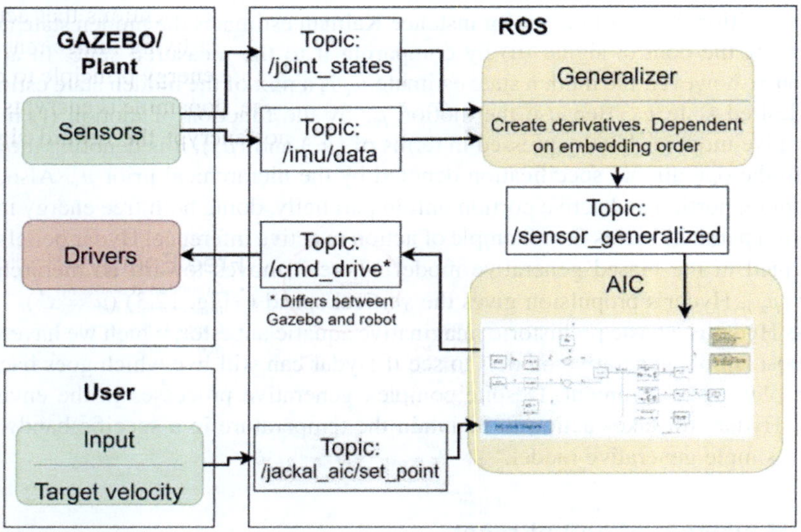

FIGURE 12.4

Component overview.

4. Artificial intelligence framework

In software companies, the concept of a simple 2 by 2 matrix is frequently used to position/compare different technologies to select the right one. If it is inspired by neurological/biological observations, or is it calculated and inspired by an abstract principle. If the capability is built on individual very particular mechanisms pop up, or if is captured by the complexities and the results of a wholesome ecosystem. Below you will find how the work of Karl Friston compares to the other AI frameworks (Fig. 12.5).

We see the free energy principle in the top right part of the quadrant. That is the leader quadrant if the dimensions are the capability to deliver and the technological vision where the dimensions vary. In this very interesting podcast,Karl Friston shares fascinating ideas on the free energy principle and talks about the two dimensions of this AI framework [17].

In the second half of the podcast, Karl Friston explains in his words the concepts of the principle, how "things" and "the-others-things-that-are-not-the -thing" can be described by the boundary and what needs to be in place to keep that boundary existing to be a separate "thing." The Free Energy is more fundamental principle for all things which is described in the second law of thermodynamics, where entropy increase over time. Things counteract this natural tendency to disorder, which lowering the Free Energy. Notice in the podcast Karl Friston mentions that action differentiates "living things" and "nonliving things" because living things react to the environment. Also, notice that Karl Friston uses the free energy principle to predict the use of biological records in the brain. For instance, dopamine is encrypts the precision of the prediction errors. Dopamine does not encrypt the reward signal

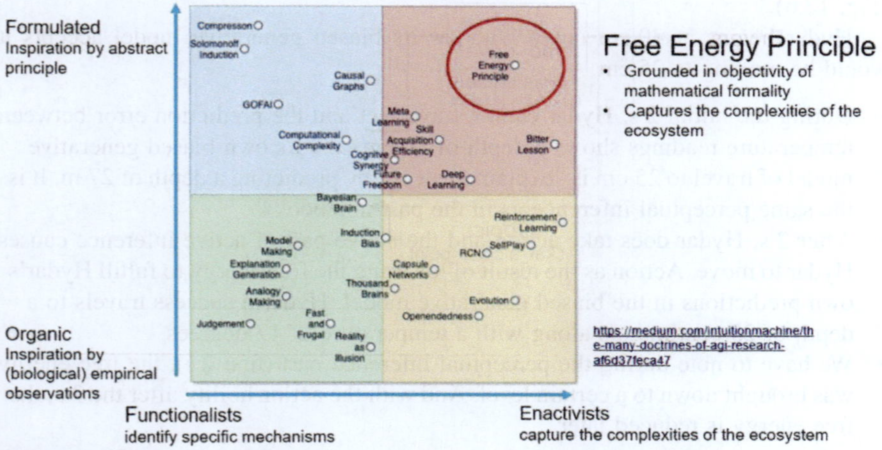

FIGURE 12.5

Map of the quest for true AI.

which is said in Reinforcement Learning but encrypts the precision of a forecasted reward. For example, a waited reward does not provide a dopamine rush.

In the example is that you would not look at the separate gas molecules if you want to understand the thermodynamics of a gas. We need not concentrate at individual neurons understanding their collective behavior. However, if we go high we would skip the basic principles of common microcircuitry, connection, and architecture. If we view the brain is performing some form of inference, the level of knowledge you need is defined on the level of population dynamics bringing the architectural principles of the brain anatomy and not the behavior of individual neurons. In the AI framework picture above therefore shows capturing the complexities of the environment.

5. Experiments conducted using Hydar

Let us visualize the principle in action and display active inference in our simulation-situated environment. In the following examples, Hydar can affect its body temperature because it has the ability to travel above and below into warmer or colder water. It will do it by reducing the free energy of its biased generative model. It is biased since Hydar believes it is traveling toward its hierarchical prior $\mu v \mu v$.

5.1 Experiment—The first

In the example, Hydar holds a depth of 30 cm and is getting chill. It climbs up to a depth of 25 degrees and affordable temperature of 17 degrees.

For example, there is no noise in the generative process yet (even though the generative model expects noise to be available) to best see how the faith develops (Fig. 12.6).

Hydar begins at 30 cm depth whereas its biased generative model accepts it would be moving to 25 cm.

- During the initial 2 s, Hydar cannot travel yet and the prediction error between temperature readings shows a depth of 30 cm, and its own biased generative model of travel to 25 cm is "explained away" by predicting a depth of 27 m. It is the same perceptual inference as in the past notebooks.
- After 2 s, Hydar does take action and the active part of active inference causes Hydar to move. Action as the result of lowering the free energy to fulfill Hydar's own predictions in the biased generative model. Hydarin success travels to a depth of 25 centimeters along with a temperature of 17 degrees.
- We have to note during the perceptual inference part (first 2 s), the free energy was brought down to a certain level. And with the action ability after the 2 s, the free energy is reduced later.

Hydar can act on the living environment! It has the capability to take action and travel to a predicted depth/temperature.

```
w covariance matrix =
[[ 0.1  0. ]
 [-0.   3.2]]
z covariance matrix =
[[ 0.1  0. ]
 [-0.   3.2]]
```

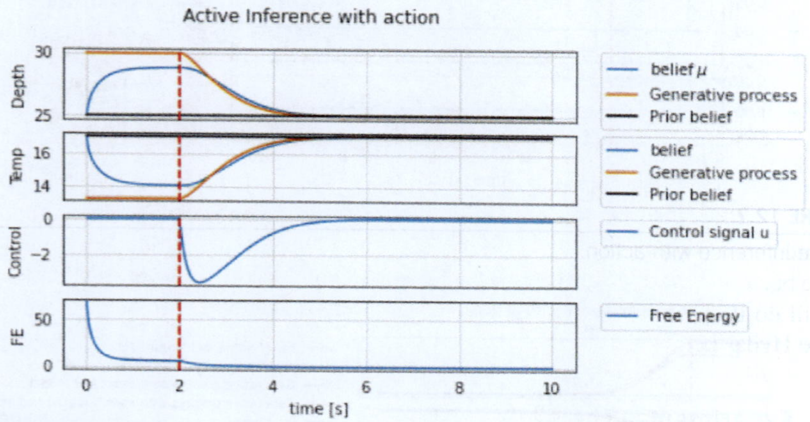

FIGURE 12.6
Active inference with action.

5.2 Experiment—Observe what is expected

When we do a quick verification, what happens if the prior prediction and real sensor measurements do match? Hydar's previous and real depth is 25 cm, so it is predicting and received a temperature of 17 degrees (Fig. 12.7).

As expected,

- the sensor measurements and previous confirmation immediately accepts, only one observation is required for Hydar to know the depth No action and The Free Energy is zero, optimal

5.3 Experiment—Different forward models

The real world in the current scenario talks about the true time-changing forward that can only be estimated. It is estimated with a constant (simple +1 or −1, the sign is really important, the direction of the forces), and the learning rate auau is used as a tuning property to get sufficiently fast action updates. This experiment compares an actual forward model versus an estimated forward model (Fig. 12.8).

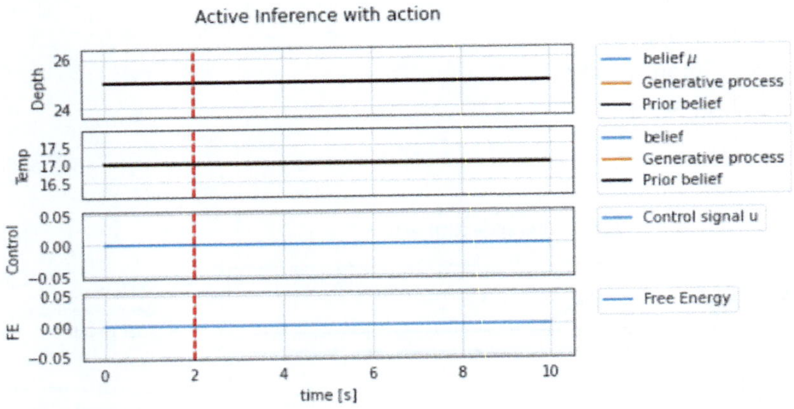

FIGURE 12.7

Active inference with action.

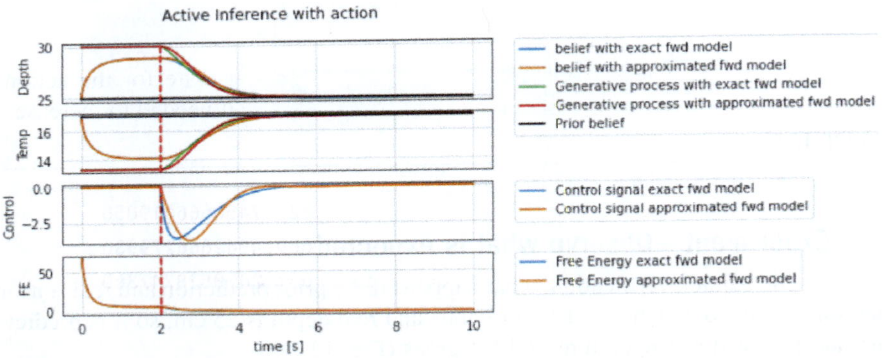

FIGURE 12.8

Active inference with action.

- As we visualize some tuning of the learning rate of u, the graphs begin to match. So the fundamental idea that the main idea of the forward model is the sign, the direction of the force needs to be right, and the rest gets tuned with the learning rate, is ensured in this experiment.
- The estimated forward model is a simple -1 used for all embedding orders. You might fight that the sign of ddg_gp is actually not -1 all places in this experiment. Where "sign" is considered as a first order approximation (move up or down), "sign+" is more complicated and comes nearer the "real" evaluations denote the sign for every embedding order is correct. In the robot arm example, putting positive torque gives the similar sign for all embedding orders. In this

example it is false because of the selected function of sensory mapping g, hence this remark.

- Also note that in the actual fwd method calculation the first entry of the fwd model is 0 because $\partial y \partial u = 0 \partial y \partial u = 0$, the prediction error on the position is unrealistic; only estimation mistakes on the higher derivatives (e.g., speed, acceleration, etc) count. In the estimated forward model it is a -1 on all embedding orders also adding the first one.

5.4 Experiment—Higher embedding orders

Same example as Experiment 1.01 but now with higher embedding orders (Fig. 12.9).

As we observe, the lines are highly comparable. To get the first idea of the differences, we evaluate the mean squared error of the three embedding orders. Embedding order 2 gives better output than embedding order 1, unexpectedly embedding order 3 gave lesser outputs than 2.

5.5 Experiment—Colored noise

With some actual colored noise in the generative process (Fig. 12.10).

- In this experiment, it becomes fast clear that the learning rate for the action (control signal) is important. As you can see in the graph, it needed to increase by a factor of 30 to get to a similar performance.

```
Mean Squared Error embedding order 1 is:  821.7456666619056
Mean Squared Error embedding order 2 is:  609.404266279936
Mean Squared Error embedding order 3 is:  687.2470528747369
```

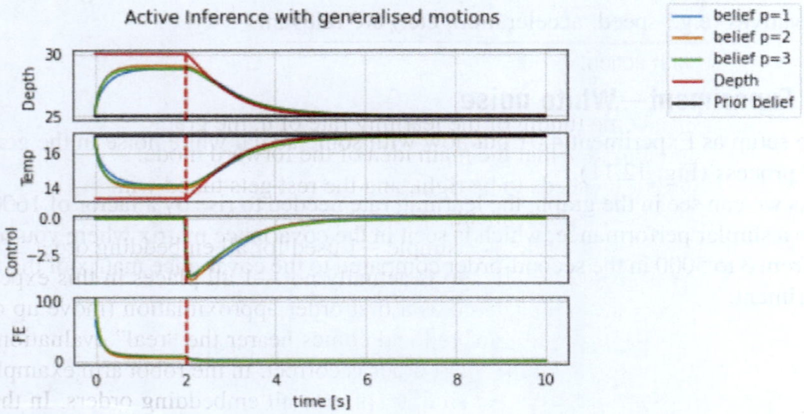

FIGURE 12.9

Active inference with generalised motions.

```
w covariance matrix =
[[ 0.1   0. ]
 [ -0.  100. ]]
z covariance matrix =
[[ 0.1   0. ]
 [-0.  100. ]]
```

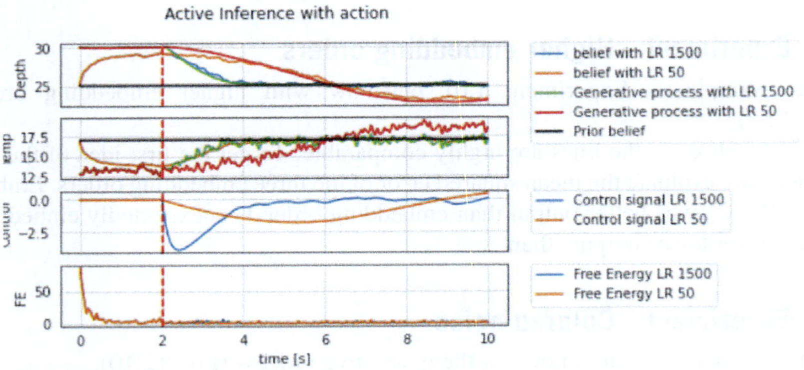

FIGURE 12.10

Active inference with action.

- The main thief for the need for a significant rise in the learning rate is the precision matrix for no-noise versus smooth-noise. For comparison, I printed the covariance matrix (inverse of the precision matrix) where you visualize a rise from 3 to 100 in the second-order comparison to the covariance matrix of the first experiment to study. In the actual fwd method calculation, the first entry of the fwd model is 0 because $\partial y \partial u = 0 \partial y \partial u = 0$, so the 0.1 in the covariance matrix is not up to the mark; only prediction errors occur on the highest derivatives (e.g., speed, acceleration, etc.) are relevant.

5.6 Experiment—White noise

Same setup as Experiment 4.01 but now with some actual white noise in the generative process (Fig. 12.11).

As we can see in the graph, the learning rate needed to rise by a factor of 1600 to get to a similar performance, which is seen in the covariance matrix where you see a rise from 3 to 5000 in the second-order compared to the covariance matrix of the first experiment.

```
w covariance matrix =
[[ 1.e-01  0.e+00]
 [-0.e+00  5.e+03]]
z covariance matrix =
[[ 1.e-01  0.e+00]
 [-0.e+00  5.e+03]]
```

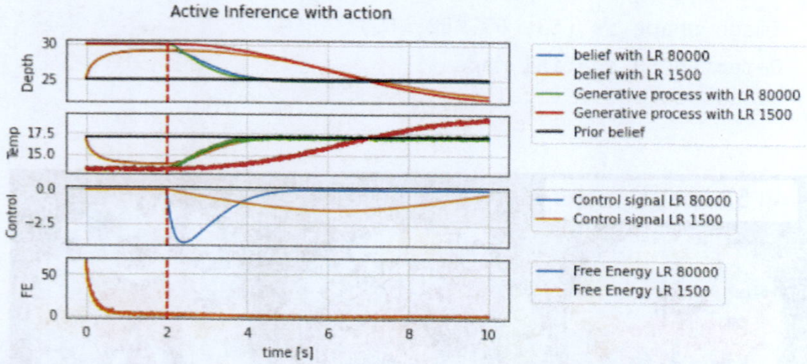

FIGURE 12.11

Active inference with action.

5.7 Plotting 4D probabilistic atlas maps

Probabilistic atlasing is a research technique in which the goal is to create anatomical outlines that keep it themselves quantitative data on cross-subject variations in brain structure. A digital probabilistic atlas of the human brain, building precise statistical information on positional variability of important functional and anatomic interfaces, will correct many current atlasing problems, as it particularly stores data on population variability (Fig. 12.12).

5.8 Plotting a statistical map

Statistical parametric mapping or SPM is a statistical method for analyzing variations in brain activity registered during functional neuroimaging experiments. The measurement methodology depends on imaging technologies (e.g., fMRI and PET). The scanner gives a map of the region that is denoted as voxels. Each voxel shows the work of a specific volume in three-dimensional space. The absolute size of a voxel varies depending on the trend. fMRI voxels formally denote a volume of 27 mm^3 (a cube with 3 mm length sides).

Parametric statistical models are considered at each voxel, by using the general linear model to describe the data variability with respect to experimental and confounding reactions, along with residual variations. Hypotheses expressed with

```
Image shape is (53, 63, 52, 53)
Detected 53 spatial maps
--------------------------------------------------
Image shape is (53, 63, 52, 53)
Detected 53 spatial maps
--------------------------------------------------
Image shape is (53, 63, 52, 53)
Detected 53 spatial maps
--------------------------------------------------
```

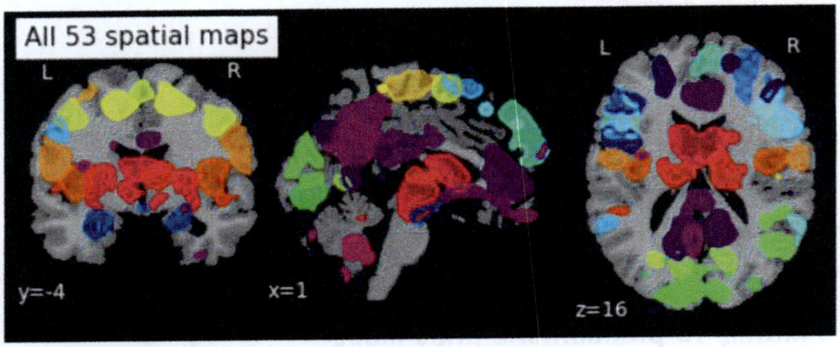

FIGURE 12.12

All 53 spatial maps.

respect to the model parameters are calculated at every voxel with univariate statistics.

Analyses may analyze variations over time (i.e., correlations between a task variable along with brain activity in a specific location) by using linear convolution models of how the calculated signal is generated by base changes in neural activity.

As many statistical experiments are carried out, adjustments are made to restrict type I errors (false positives) potentially resulting by the comparing levels of activity over many voxels. A type I error would result in not true assessing background brain activity in relation to the task. Adjustments are made based on the number of resels in the image and the theory of continuous randomly generated fields for setting a new methodology for statistical importance that could adjust for the complication of many comparisons (Fig. 12.13).

5.9 Glass brain visualization

Glass brain is an instrument that connects the electrical activity of our brain with real-time results. The anatomically real-time 3D brain will depict real data from

```
Image shape is (53, 63, 52, 53)
Detected 53 spatial maps
(53, 63, 52)
------------------------------------------------
Image shape is (53, 63, 52, 53)
Detected 53 spatial maps
(53, 63, 52)
------------------------------------------------
Image shape is (53, 63, 52, 53)
Detected 53 spatial maps
(53, 63, 52)
------------------------------------------------
```

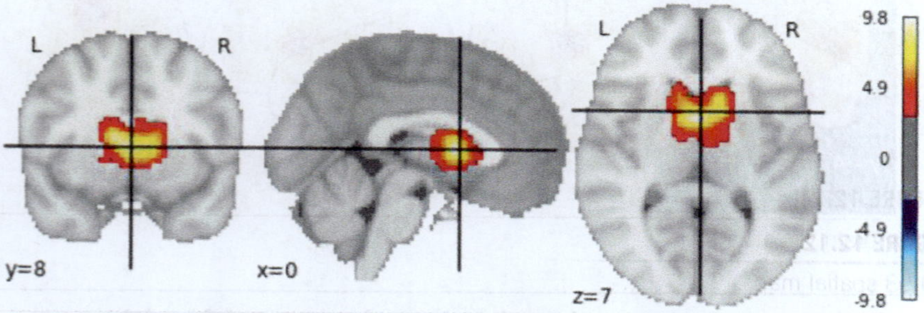

FIGURE 12.13

Plotting a statistical map.

electroencephalographic (EEG) signals taken from an ultra-moduled and designed EEG cap. These data are connected to the source of that electrical activity, that is, the specific location of the brain. The base brain model is created through MRI scans so that the EEG data are accurately connected to an individual's brain model.

Various colors are given to the varying signal frequency bands to generate a beautiful interactive artwork that crackles with energy, depicting how data are transported (or at least estimated to do so) between various regions of the brain (Fig. 12.14).

```
Image shape is (53, 63, 52, 53)
Detected 53 spatial maps
(53, 63, 52)
------------------------------------------------
Image shape is (53, 63, 52, 53)
Detected 53 spatial maps
(53, 63, 52)
------------------------------------------------
Image shape is (53, 63, 52, 53)
Detected 53 spatial maps
(53, 63, 52)
------------------------------------------------
```

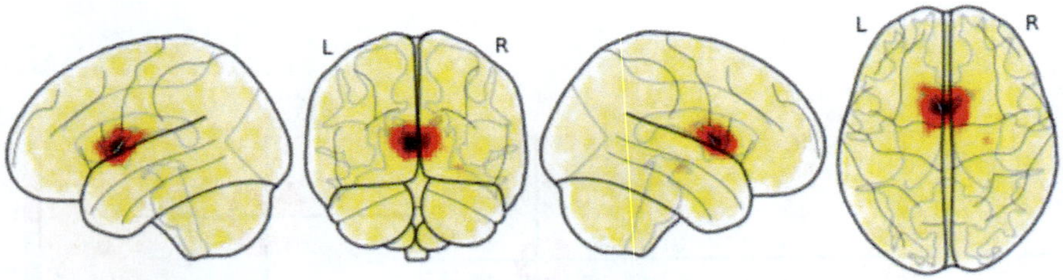

FIGURE 12.14

Glass brain visualization.

6. Survey on impact of technology in mental health

6.1 Have/had mental health disorder

Forty-one percent of the respondents ensure that they have at the instant a mental health disorder, while more than half of the respondents said that they held a mental health disorder in the past [18]. What bothers is the amount of respondents who answered "maybe" to this question, meaning a necessity to guide, identify, and address any issue that raised that "maybe" response in the first place (Figs. 12.15 and 12.16).

6.2 Companies taking mental health seriously

From the mean value in the survey, we would say yes, but not as much as it would be necessary. More than half of the responses have a Mental Health Coverage result, but not the data of finding more information about this, meaning that most companies talk about mental health, but will not implicate themselves actively in it. More than 70% of respondents NOT know would be protected in the case of talking to a senior person about their mental health problems, while half part of them are

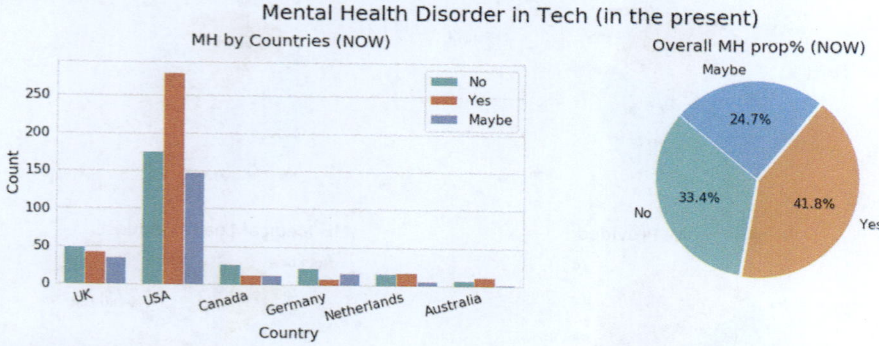

FIGURE 12.15

Mental health disorder in technology present.

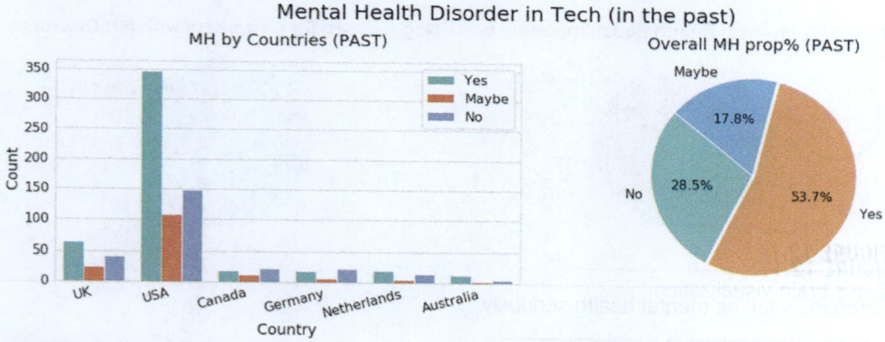

FIGURE 12.16

Mental health disorder in technology present.

not sure that the company grants a medical leave. Finally, most responders do not know if their company is taking mental health problems the same as physical health. What encourages is that 94% of respondents do not think that there result negative consequences for coworkers who face mental health problems (Fig. 12.17).

6.3 Discussing mental health at work

Sixty-seven percent of respondents do not know that there were negative consequences to talking about mental health with the employer. There is also a big community of people that do not know if they would be mentally free of talking with their employer about mental health. Alarming is the percentage of people who actually know that coworkers will view them badly after confessing to having a mental health problem. Nearly half of the respondents are mentally free to talk about mental health with their senior faculty (Fig. 12.18).

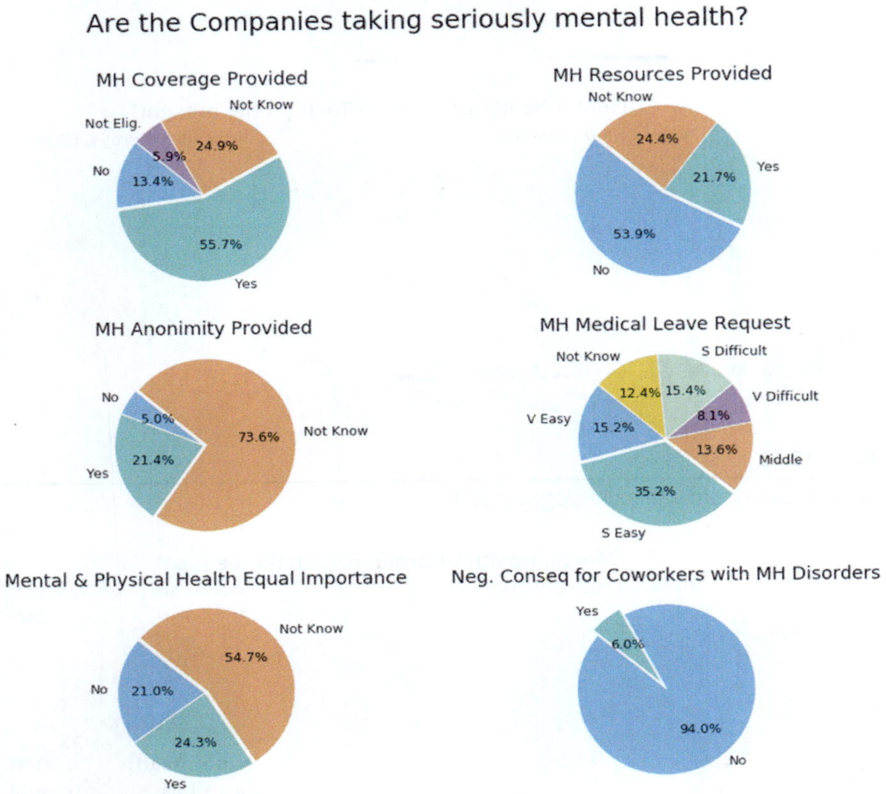

FIGURE 12.17

Companies taking mental health seriously.

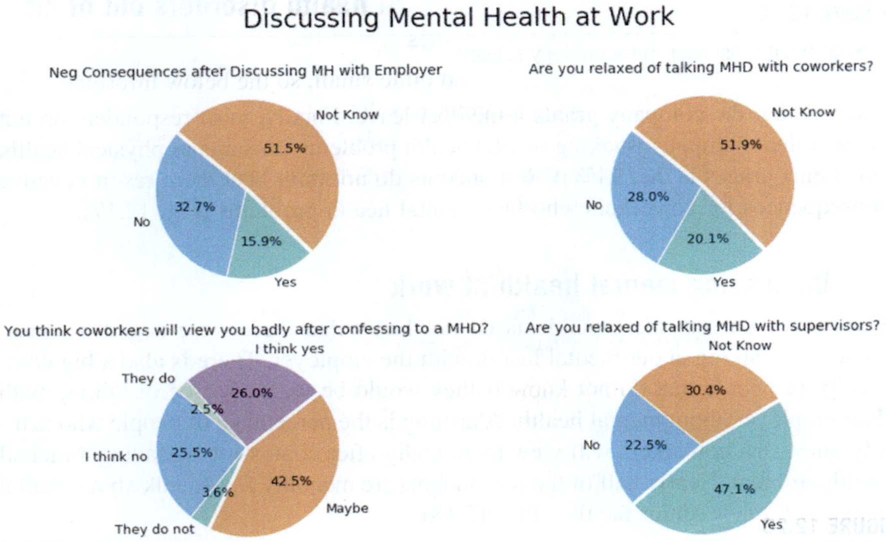

FIGURE 12.18

Discussing mental health at work.

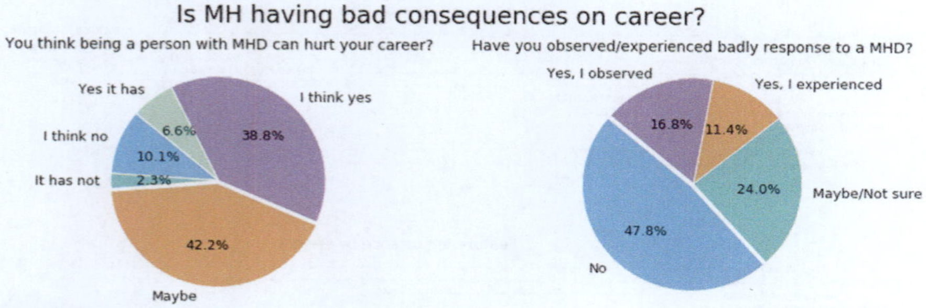

FIGURE 12.19

Mental health having consequences on career.

6.4 Mental health having consequences on career

Most people think mental health disorder would spoil their career, but more than half of the responders did not experience a bad response to mental health situation around them. These questions still show biases about mental health and its result on someone's life (Fig. 12.19).

6.5 Being open about mental health with friends and family

Sixty-three percent of respondents are open about talking mental health with their friends and family. Almost 20% of responders are not willing to talk about mental health with closer people (Fig. 12.20).

6.6 Percentages of people with mental health disorders out of all respondents in the United States

The data for each state are changeable and quite small, so the below information in the map should be considered with a grain of salt.

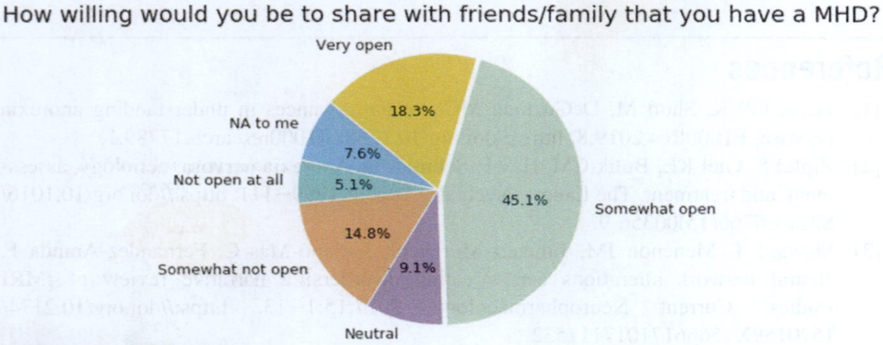

FIGURE 12.20

Being open about mental health with friends and family.

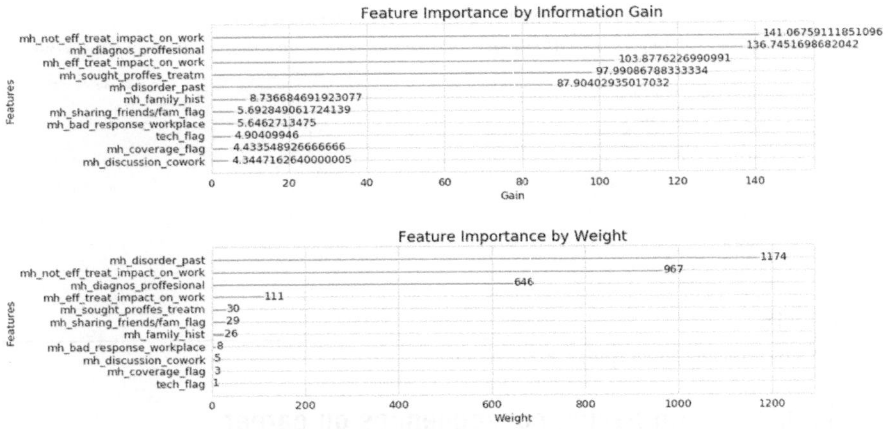

FIGURE 12.21

Percentage of people with mental health disorder in the United States.

Percentage of people with MHD in every state (from total respondents) (Fig. 12.21).

7. Conclusion

In our point of view, the openness of companies and people with respect to mental health in tech is a bit powerful than I would have anticipated, but there is improvement needed. The number of people who are certain of future negative results, in the workplace or at home after talking about mental health, should be ignored. There are some biased thoughts among the respondents, mental health disorders still being viewed with a strict mentality, when in reality people pay more attention and accepted them.

References

[1] Frank GWK, Shott M, DeGuzman MC. Recent advances in understanding anorexia nervosa. F1000Res 2019;8. https://doi.org/10.12688/f1000research.17789.1.

[2] Zipfel S, Giel KE, Bulik CM, Hay P, Schmidt U. Anorexia nervosa: aetiology, assessment, and treatment. The Lancet Psychiatry 2015;2:1099–111. https://doi.org/10.1016/S2215-0366(15)00356-9.

[3] Steward T, Menchon JM, Jimenez-Murcia S, Soriano-Mas C, Fernandez-Aranda F. Neural network alterations across eating disorders: a narrative review of fMRI studies. Current Neuropharmacology 2017;15:1–13. https://doi.org/10.2174/1570159X15666171017111532.

[4] Meneguzzo P, Collantoni E, Solmi M, Tenconi E, Favaro A. Anorexia nervosa and difusion weighted imaging: an open methodological question raised by a systematic review and a fractional anisotropy anatomical likelihood estimation meta-analysis. International Journal of Eating Disorders 2019;52:1237−50. https://doi.org/10.1002/eat.23160.

[5] Bullmore E, Sporns O. Complex brain networks: graph theoretical analysis of structural and functional systems. Nat Publ Gr 2009;10:186−98. https://doi.org/10.1038/nrn2575.

[6] Rubinov M, Sporns O. Complex network measures of brain connectivity: uses and interpretations. NeuroImage 2010;52:1059−69. https://doi.org/10.1016/j.neuroimage.2009.10.003.

[7] Boccaletti S, Latora V, Moreno Y, Chavez M, Hwang DU. Complex networks: structure and dynamics. Physics Reports 2006;424:175−308.

[8] Bullmore E, Sporns O. The economy of brain network organization. Nature Reviews Neuroscience 2012;13:336−49.

[9] Bassett DS, Bullmore E. Small-world brain networks. The Neuroscientist 2006;12:512−23.

[10] van den Heuvel MP, Sporns O. Network hubs in the human brain. Trends Cogn Sci 2013;17:683−96. https://doi.org/10.1016/j.tics.2013.09.012.

[11] Favaro A. Brain development and neurocircuit modeling are the interface between genetic/environmental risk factors and eating disorders. A commentary on keel and forney and friederich et al. International Journal of Eating Disorders 2013;46:443−6.

[12] Plitman E, Patel R, Mallar Chakravarty M. Seeing the bigger picture: multimodal neuroimaging to investigate neuropsychiatric illnesses. Journal of Psychiatry & Neuroscience 2020;45:147−9.

[13] Frank GWK, Favaro A, Marsh R, Ehrlich S, Lawson EA. Toward valid and reliable brain imaging results in eating disorders. International Journal of Eating Disorders 2018;51:250−61. https://doi.org/10.1002/eat.22829.

[14] Zalesky A, Fornito A, Bullmore ET. Network-based statistic: identifying diferences in brain networks. NeuroImage 2010;53:1197−207. https://doi.org/10.1016/j.neuroimage.2010.06.041.

[15] Learn by example: active inference in the brain -5 Webpage · 2021 · Charel.

[16] Learn by example: active inference in the brain -1 Webpage · 2021 · Charel.

[17] Karl Friston: neuroscience and the free energy principle | Lex Fridman Podcast #99 Webpage · 2020 · Lex Fridman.

[18] Model and visualize mental health in tech Webpage · 2020 · Andradaolteanu.

Index

'*Note:* Page numbers followed by "f" indicate figures, "t" indicates tables.'

K

L

M

N

O

P

Q

R

S

T